Emer~~~~~~~~~~

Emergency
Crisis on the Flight Deck
2nd edition

Stanley Stewart

Airlife

'One thorn of experience is worth a whole wilderness of warning.'
James Russel Lowell

This 2nd edition published by Airlife Publishing Ltd 2002
First published in the UK in 1989 (hardback) and 1992 (paperback)
by Airlife Publishing Ltd

British Library Cataloguing in Publication Data
A catalogue record for this book is available from the British Library.

ISBN 1 84037 393 8

Typeset by Phoenix Typesetting, Burley-in-Wharfedale, West Yorkshire
Printed in England by MPG Books Ltd., Bodmin, Cornwall

Contact us for a free catalogue that describes the complete range of Airlife books for
aviation enthusiasts.

Airlife Publishing Ltd
101 Longden Road, Shrewsbury, SY3 9EB, England
E-mail: sales@airlifebooks.com
Website: www.airlifebooks.com

Contents

Acknowledgements

Much help, assistance and advice was received from many people during the writing of this book and the author is deeply indebted to all those who so kindly contributed. Without the generous support of those involved in the incidents, and others in the aviation industry, this book would not have been possible. To all those who so kindly helped, the author would like to express his heartfelt thanks. **Forced Entry**: Captain Paul Whetham and Captain Jeff Morgan; **That Falling Feeling**: Captain Tim Lancaster and Captain Alastair Atchison; **Pacific Search**: Captain Gordon Vette and Captain Jay Prochnow; **The Windsor Incident**: Captain Bryce McCormick; **Don't be Fuelish**: Captain Bob Pearson and Captain Maurice Quintal; **The Blackest Day**: Captain Pat Levix; **Ice Cool**: Captain Tom Hart; **Roll Out the Barrel**: Captain Harvey 'Hoot' Gibson and Mr Harold F. Marthinsen, Director ALPA Accident Investigation Department; **Strange Encounter**: Captain Eric Moody, Captain Roger Greaves, Senior Engineer Officer Barry Townley-Freeman, with assistance from Captain Frank Avery (747 Instructor).

Any errors remaining are, of course, entirely my own.

Introduction

Flying is one of the safest forms of modern transport and, as a means of travelling quickly over long distances, its role is unsurpassed. The passenger of today is transported with amazing ease from one side of the globe to the other, with the problems of the earth being left far below. For present-day flight crew, however, it is another world, with the view from the flight deck offering a different sight. Aircrew are only too well aware of the hostile nature of their working environment and, armed with such knowledge, are both ready and able to overcome the difficulties.

Aircraft frequently cross great empty oceans, vast featureless deserts, immense ice wastelands and enormous desolate regions in complete safety in spite of the adverse conditions of the terrain below. Aircrew are trained for all contingencies, and survival equipment for sea, ice and desert is carried aboard. On a journey half way round the world many areas of conflict may be crossed without a single sound of the calamity below being heard, the loudest report in the cabin being the pop of a champagne cork. Such danger zones can be traversed or circumvented in safety when approached with vigilance and care.

Natural disasters, civil strife and famine also prove of little effect for the high-flying traveller, although for staff and crew at transit stations the problems may be enormous. In the skies of starving Africa, passengers sitting miles above the horror indulge their tastes in international cuisine. Cosseted by eager airlines, the modern traveller is borne, with diligence and care, in genuine security over the trouble spots of the earth. Journeys over half the world are now so commonplace and scheduled arrivals so frequent that delays of an hour or so can annoy passengers. To arrive, say, a few hours late in Auckland after a 12,000 mile trip from London can be quite unacceptable to some. Yet if the same passengers stopped only for one moment to think of such a journey, they could not fail to be impressed by the accomplishment. The world may be shrinking, but it is not quite as small as we are led to believe. It is still a hazardous place for the unwary and even short flights are rarely as simple as they seem.

In spite of advanced technology and the magic of computers, the movement of something as big as a Boeing 747 from one side of the world to the other is an operation of complex proportions. Anyone who has contemplated driving their automobile in a foreign country will immediately recognise the problems; the difficulties in operating a big jet worldwide are immense; overflying rights, landing permission, insurance arrangements, fuel payments, cargo and passenger quotas are but a few of the challenges, all of which have to be negotiated between governments and often between countries barely talking to each other. The capital equipment required to service a large international airline operation is enormous: aircraft, offices, sales shops, hangars, terminal buildings and a plethora of expensive vehicles from mobile steps to push-back trucks. With air fares over long distances still comparatively low in respect of present-day incomes, it is a wonder that airlines make any profit at all. That they do and, in some cases, manage quite handsome returns in the face of fierce competition, is a great credit to the managers who run these large and costly outfits.

The fact that airlines function successfully throughout the globe is due to the dedicated and hardworking people within the industry who make the system work, in spite of the problems. Managers, office staff, sales persons, accountants, engineers, maintenance personnel, traffic supervisors, the backroom people of the world's airlines, all make a significant contribution, as well as the aircrews at the sharp end of the operation. For passengers, the smooth, comfortable and effortless transition from one place to another is not accomplished without a great deal of exertion from all concerned. Much of the effort is, of course, unobserved by the travelling public, not least the skills of the flight crew in flying the aircraft from departure to destination. It is an esoteric world where few of even the most well travelled of passengers have been permitted to enter.

The airline pilot's job today is essentially one of operations director and systems manager, but even on the most sophisticated of electronic flight decks the human contribution is significant. In spite of the advances in computers and electronics, machines can do no reasoning for themselves and creative thinking is still a necessary facet of the job. Aircraft computers can do only so much and modern electronic capabilities are not quite as fantastic as the public is led to believe. Malfunctions do occur and there are many traps for the unwary. Flight crews, of course, are alert to the problems, but are sometimes too eager to assure passengers by telling them how easy it has all become. Basic airmanship (i.e. the collective practical application of training, skill, experience and professional judgement) is still required to be exercised by all flight crews at all times.

All aircraft computers are required to be programmed before flight for each journey and the autopilot, when engaged, has to be instructed on every move. Automatic guidance of a 747, for example, down a narrow radio beam to accomplish an automatic landing in almost blind, foggy conditions with 400 people on board is not a task to be taken lightly. The automatics have to be very carefully monitored for malfunctions and the autopilot has to be told what to do at each stage of the approach. In the fog, of course, the wind is calm and the air still, and in such circumstances automatic landings can be effective. When the wind is blustery and conditions bumpy, however, especially with a strong crosswind on landing, the autopilot cannot cope and the pilot has to take over and land the aircraft.

On most flights, crew operation is routine with standard procedures being followed but, of course, circumstances do change, even when flying repeatedly on the same route. Take-off and landing delays, work at airports, equipment malfunctions, re-routeings, adverse weather and so on, all present difficulties. Flight crews, however, not only have to perform the routine well, a highly skilled procedure in itself, but have to cope with any emergency which may arise. When severe weather strikes, systems malfunction, engines fail or aircraft fires erupt, it is sometimes only the skill of the flight crew which lies between safety and disaster. Flight crews have at all times to be alert to every situation. When a major emergency occurs decisions are made, sometimes in a split second, which can affect the safety of the aircraft and perhaps many hundreds of lives. Here the captain comes into his own and the training of the flight crew is put to the test.

Airline crews are well-trained, highly motivated and dedicated professionals. Although mistakes are sometimes made and accidents occasionally happen, the high level of safety evident in the airline industry is a testament to the excellent standard maintained by all concerned. Too much publicity these days is given to the rare demise of an aircraft, with little being known of the incidents which, owing to aircrew professionalism, end safely and well. Many of the events told within this book are known in aviation circles, with only a few outside, and most of those involved remain unsung heroes. *Emergency* outlines a number of dramatic incidents and reveals crew procedures during the difficulties, inviting the reader into the exclusive environment of the flight deck to observe the operations.

The author, Stan Stewart, has flown for over thirty-five years and has operated heavy jets in British Airways for thirty years. A graduate engineer, 747 pilot and aviation writer, he is uniquely qualified to write on aviation matters. *Emergency* is a celebration of the skills and abilities of pilots and flight engineers throughout the world, and the following chapters can be left to speak for themselves.

Chapter 1

Forced Entry

On 11 September 2001, the world of aviation changed for ever. In the years before the events of that day, when the twin towers of the World Trade Centre in New York were attacked, the aviation and security industries were justifiably proud of their record in preventing the unlawful seizure of aircraft. Although lapses had occurred and forced entry to flight decks had been recorded, the efforts of the security industry in denying terrorists access to aircraft and in preventing weapons from getting aboard, and the determination of individual states to prosecute and punish culprits, seemed to be paying dividends. Unknown to aviation at the time, however, the industry had been lulled into a false sense of security and, on that fateful day in September, everything changed.

On the morning of 11 September 2001, nineteen individuals, all of whom were bent on suicide, simultaneously seized four aircraft. To the horror of the world and the total dismay of the aviation and security industries, the Twin Towers in Manhattan were destroyed, the Pentagon in Washington DC was badly damaged and 3,000 people lost their lives. It was a monumental crime against humanity and an aviation tragedy of unprecedented proportions.

In the decade before 11 September, hijackings were rare, as not only did the crime seem to be out of vogue, but also havens for hijackers had diminished considerably as Third World governments had become less tolerant of such behaviour. Security at airports had been tightened extensively and metal detectors, the searching and X-raying of hand and hold baggage, and the identifying of individual suitcases had all helped to improve safety, not only with regard to hijackings, but to bomb scares as well. The authorities also seemed much more able to cope with terrorists and the risk of imprisonment for such criminals was high. On occasions, culprits were even slain. The action of authorities and the implementation of security measures appeared to be effective but, at the time, no one imagined that any group of perpetrators would be prepared to commit mass suicide and cause death and destruction on such a massive scale.

Prior to 11 September, individual countries, and even airlines within the same country, did not conform to a unified approach to security but adopted various attitudes in combating terrorism in the air. In the United States, for example, the Federal Aviation Administration (FAA) forbade

1

passengers' visits to the flight deck during flight for all airlines. The cockpit door was locked at all times and on some flights armed sky marshals were carried. In some other countries, crews were themselves encouraged to physically resist hijacking attempts.

The laws of most European nations, and procedures on their airlines, permitted a more relaxed attitude to flight deck access. It was believed, not unreasonably, that determined terrorists armed with anything from a knife to a small firearm could easily overpower an unarmed crew by force, and that the line of least resistance was the safest approach for passengers. Hijackings, in fact, although sometimes bloody and violent, had rarely resulted in an aircraft accident and experience had shown that once overpowered, crews complying with terrorists' instructions could best maintain safety. The locked door policy and the barring of passengers to the cockpit in the air on US aircraft had never been a totally effective deterrent in stopping forced entry to the flight deck or in preventing the unlawful seizure of an aircraft by determined assailants, as evidenced by the events of 11 September. In addition, in the pre-11 September era, it was not unusual for terrorists to threaten to attack passengers if the flight crew refused to unlock the cockpit door. It was a very callous captain indeed who could listen to passengers being injured, or even killed, outside the door while continuing to deny access to the flight deck, especially since the evidence indicated that the line of least resistance was the safest. The evidence also indicated that once terrorists seized control the likely outcome was damage being caused only to the aircraft and that, in almost all cases, most, if not all, passengers would eventually be freed. Since 11 September, of course, the restricting of access to the flight deck has become of paramount importance and every effort is being made to prevent forced entry by the installation of secure cockpit doors.

The strengthening of flight deck doors and the policy of keeping them securely locked in flight is now a key feature in preventing any repetition of the 11 September atrocity. The design of the new secure doors varies slightly, but they are mostly Kevlar armoured and are locked by a cross-bar locking device of which the main component is a 19 mm (0.75 in) steel bar that can withstand a 680 kg (1,500 lb) force. The door itself must also be able to withstand the ramming force of a fully laden food trolley weighing 136 kg (300 lb). Almost any security system can eventually be defeated by a determined attacker, of course, but the resistance of the door is designed to give the flight crew time to land, or to adopt other procedures such as violent movement of the controls to throw the attackers off balance, or depressurising the cabin to render the assailants unconscious. The doors, however, are not designed to contain the impact of a bomb, as it is assumed that an explosive device large enough to blow any secure barrier would probably destroy the aircraft anyway. The door and bulkhead are also only ballistic resistant and not bullet proof, as heightened security at airports is expected to prevent firearms being boarded, as it did on 11 September.

The Israelis lead the world in aviation security, and access to the flight decks on El Al, for example, is via two reinforced doors with a sterile area between. The arrangement prevents an attacker rushing the cockpit when one of the doors is open, and the system is being studied for longer-term security solutions. In the meantime, the programme is continuing with a single secure door system and the Federal Aviation Administration (FAA) requires completion on US aircraft by April 2003. The retrofit is a daunting task and, as other nations follow suit, the number of commercial aircraft involved could be as many as 10,000 worldwide, with the estimated cost being $2 billion.

The installation of secure cockpit doors, however, creates as many safety problems as it solves security requirements. To begin with, crew co-operation in an emergency is vital, and a firmly locked door between flight and cabin crews does nothing for crew communications. A further problem is the occurrence of a sudden depressurisation. A locked door forming a secure barrier system could create dangerous loads on the structure if unequal pressures resulted on either side of the door and bulkhead. Secure cockpit doors are, therefore, provided with hinged panels and vents arranged to be opened by quick-release latches, triggered by a pressure rate sensor, if an explosive decompression occurs. The integral strength of the door, however, is maintained with the latches in place and anti-jam panels are also fitted to allow trapped flight crew to escape from the cockpit in the event of an accident. The doors can be opened from the outside by the operation of locks via numeric keypads that can be over-ridden by the pilots, and some door systems can also be locked and unlocked remotely from the cockpit centre-pedestal.

Another feature being introduced to improve security in the air is a video camera surveillance system for monitoring cabin activity, whereby the behaviour of passengers can be viewed from a display on the flight deck. The use of closed-circuit television (CCTV) is widespread in the security industry at large, of course, and has now become as important an issue as secure cockpit doors for security in the air. There is little doubt it will become a mandatory aviation item. Surveillance equipment produced to date includes systems with a minimum of four cameras, effective for both day and night use, and a touchscreen monitor on the flight deck that can display up to four video images. Also integral to the system is a crew alert wireless pager that can be activated by a flight attendant pressing a 'panic button' to silently alert the pilots of an emergency in the cabin. The locations of the four cameras are two displaying different views outside the flight deck door, one above and one just aft in the ceiling, giving sight of a 2.4 m (8 ft) approach area to the cockpit, a third in the forward galley and a fourth at the front of the cabin looking rearwards down the aisle. Unlocking of the flight deck door from the outside using the numeric pads produces an aural and visual warning that alerts the pilots to check the identity of the intruder on the cockpit monitor.

Air marshals have been in use in the US for some time although their numbers have been small and their duties restricted mostly to international flights where the risk factor was considered high. Elsewhere there has been a reluctance to employ sky marshals but, on airlines such as El Al, their use has been extensive. Attitudes changed after 11 September, especially in the United States, and sky marshals are now carried on flights as deemed necessary, and are required to be on board all flights operating into and out of Washington National Airport, owing to its proximity to so many US government buildings. In the event of a confrontation with attackers, any shooting in flight, fortunately, will not pose the same risks as previously. Sky marshals' weapons will not be loaded with standard ammunition, but will fire a compressed, powdered metal bullet that will disintegrate on impact with a hard surface such as the fuselage. It will, however, still penetrate soft skin.

Other national governments are more sceptical of licensing the use of armed sky marshals and are doubtful of their effectiveness. They also fear the risk of a law-abiding passenger being shot. Marshals will require to be constantly vigilant while remaining incognito, and the task is considered difficult. It is also argued that terrorists who are sufficiently clever to get themselves and weapons past a sophisticated airport security system will be clever enough to deal with any sky marshals on board. Many actions are available to a determined team of terrorists to flush out and identify on-board marshals. Sky marshals could, of course, offer a very useful deterrent against a group of unarmed hijackers. A team of big, powerful terrorists trained in unarmed combat would not need weapons to create havoc on an aircraft. Assistance from marshals would also prove useful in a violent air rage incident or with a violent mentally disturbed person, especially if a large, strong person was involved.

There is also a proposal in the United States to arm pilots, and the American Airline Pilots Association (ALPA) supports the plan. The pilots' firearms, however, would only ever be considered as a last line of defence. Weapons would only be used in the event of a total failure of the security system, with sky marshals being overwhelmed and attackers smashing through the flight deck door. The ensuing shoot-out might be akin to the gunfight at the OK Corral but would be preferable to a repeat of an 11 September scenario.

Secure flight deck doors, cabin-surveillance cameras and sky marshals are all second-level defence systems that assume a breakdown of security at airports, and the real task is to prevent all undesirable individuals of any kind and weapons of any description getting aboard an aircraft in the first place. Whatever the approach in the air, therefore, the answer lies in strict security precautions at airports.

Sophisticated security arrangements have been in place in Europe for many years. As far back as 1988, the catalyst for the upgrading of airport security systems throughout the continent was the destruction of Pan American Flight 103 over Lockerbie in Scotland. Clearly, the bomb had been planted at a European airport. A comprehensive security infrastructure emerged in which metal detectors, the X-raying and searching of hand baggage and the identifying and scanning of hold baggage became commonplace. In the United States the X-raying and search of passengers and carry-on baggage were effective, but the reconciliation and X-raying of hold luggage was not a feature on domestic flights. Since 11 September, airport security throughout the world has been upgraded, and police and military personnel at many international airports now assist security staff. Hand luggage is restricted and any sharp objects such as penknives, scissors and nail files are not permitted on board. Even eyelash curlers are prohibited. The terrorists who hijacked flights on 11 September carried on board carpet cutters and were able to inflict serious injury. These small knives had light plastic handles with very sharp blades and were difficult to detect. Now any item that is possible to fashion into a sharp object is banned. The 100 per cent reconciliation of luggage with every passenger and the scanning of all hold baggage before loading is also the ultimate aim

The equipment and procedures required for the detection and prevention of weapons of all kinds being boarded are in place at airports and the screening of passengers themselves is now a priority. Any determined assailant, or potentially violent individual, armed or unarmed, poses a threat, and suspects must be prevented from boarding an aircraft in the first place. One solution is the screening and profiling of passengers. The aim of screening techniques now being introduced, however, is not only to reassure the travelling public by being visible and effective, but also to prevent delays by being as quick and as unobtrusive as possible. Advanced screening systems offer identification of individual passengers by biometric techniques. Sophisticated computers can identify passengers by means, for example, of finger and palm prints, and by iris, voice and face recognition. As a result, passengers can easily be traced and identified from reservation, through check-in to boarding, and the procedure can also be used by immigration and emigration services. Databases maintained by security agencies can also be searched for behavioural characteristics or criminal records, depending on the privacy laws of the country. Even information as simple as an address, travel to certain destinations or the method of ticket payment can target an individual for additional inspection. Clearly some countries are considered suspect and a one-way ticket paid for by cash can raise concern. The small number of travellers considered as possible threats, plus some others selected at random, can be subjected to further security measures. Once again the Israelis lead the world in the profiling process, and at Ben Gurion Airport in Tel Aviv, for example, passengers are questioned before boarding. Highly trained security officers conduct interviews and can select passengers for additional security checks. Failure to answer questions convincingly and the displaying of stress-related behavioural patterns and body movements can alert staff to a possible suspect. If the security officer is not satisfied, the suspect is cross-examined by a senior officer and their luggage searched. Ultimately, of course, a suspicious individual can be denied boarding. Passenger profiling is known to be effective but, in some countries, certain aspects are in breach of civil rights and the process has to be adapted to conform to national laws.

Effective profiling of passengers not only assists in detecting terrorists but can also help to deny boarding to other undesirables such as the potentially violent. The incidence of air rage, for example, is increasing alarmingly. Air rage incidents tend to range from non-violent scuffles between unruly passengers to vicious attacks on crew. One of the worst events aboard a UK aircraft in recent years occurred in October 1998 on an Airtours charter flight from London Gatwick to Malaga in the south of Spain. A passenger, who had been drinking heavily in the terminal, was permitted to board, and it was clear to the crew when he arrived at the aircraft that he was drunk. The captain was informed and the drunken passenger was told to sit at the back of the aircraft and was refused alcohol. The drunk, however, was abusive and disruptive on the journey and the police were asked to meet the aircraft in Malaga. On arrival, whilst waiting for the officers to board, the troublemaker attacked a stewardess with an empty vodka bottle. The drunk smashed the bottle on the left side of the flight attendant's head and continued to attack her by jabbing the broken bottle into her face, then into her arm and back as she tried to protect herself. The assailant was restrained by passengers but not before the stewardess was left permanently scarred. The violent drunk, who should never have been permitted to board in the first place, was sentenced by a Spanish court to four years in prison.

On occasions, the actions of the emotionally disturbed or mentally ill passenger have also been known to disrupt flights. Such disturbances normally follow similar patterns to those of air rage, and range from the non-violent, but upsetting, to the extremely violent. Events involving the emotionally disturbed and mentally ill frequently occur without warning and often when the unfortunates are travelling alone and the crew have not been informed that a mentally unstable person is on board. Disturbances from air rage and the mentally unstable normally involve violence being restricted to the cabin, with flight attendants in the front line. On the very rare occasion, however, the actions of the emotionally

6

disturbed or mentally ill have involved flight crew and have been extremely violent. As a result of an incident in 1987, the Federal Aviation Administration (FAA) in America ordered all cockpit doors of US aircraft to be locked in flight. An armed male passenger, who turned out to be a disgruntled former employee, intruded into the cockpit of a Pacific Southwest flight over California and shot dead both the captain and the co-pilot. The aircraft crashed, killing all on board. In May 2000, a mentally unbalanced passenger aboard an All Nippon Airways Boeing 747-400 burst on to the flight deck with a knife and fatally stabbed the captain. The assailant was able to carry a knife on board owing to a security loophole, now closed, at Haneda domestic airport in Tokyo. After the incident, airlines in Japan ordered flight deck doors to be locked in flight and banned visits to the cockpit. Although both the above attacks on flight crew had been extremely violent, on each occasion the incidents were considered sufficiently isolated not to warrant responses on an international scale.

Prior to 11 September, no one imagined that the strategy of terrorists would change from hijacking passenger aircraft for bargaining purposes to the seizing of aircraft to use as weapons of mass destruction. Security measures now in place at airports and on board aircraft can prevent a recurrence of such an event and can help deny boarding to all undesirables or, if they do breach the first level of security and board, can contain their destructive actions. The new procedures will also prevent a recurrence of the violent attacks on flight crew such as those that occurred in 1987 and May 2000. Had such measures been in place in December 2000, the system would also have prevented a large, strong, mentally unstable man boarding an aircraft, forcing an entry to the flight deck and causing unbelievable havoc.

Thursday 28 December 2000 was not a typical grey, winter's day in London, but it was still cold. The wind from the west was brisk and the temperature a chilly 3°C (37°F). Rain and sleety showers had been evident earlier in the day from a partly clouded sky but some sunshine had appeared in the afternoon. A thin sliver of the new moon had risen in the morning and set during the early evening, so by late evening the sky was black. As passengers arrived at London Gatwick's North Terminal for the late evening departure of British Airways flight BA 2069 to Nairobi, the temperature had dropped to near zero, but the conditions were still dry. Many passengers were leaving London for a winter break in Africa and they may have thought longingly of the weather at their destination, where the temperature in Nairobi that day had been 27°C (81°F).

The British Airways 22:20 departure to Nairobi was one of the last flights of the day to leave the North Terminal. With most aircraft departed, the terminal was quiet and many passengers awaiting the flight mingled in the near empty building close to Departure Gate 54, or in the shops still open in the main concourse. Others more fortunate waited comfortably in the First and Club Class lounges, amongst whom were some well-known and distinguished guests. Bryan Ferry, the rock star, with his wife and teenage son were amongst them. The Ferry family was on the way to a vacation in Zanzibar and was flying to Nairobi on the first leg of the journey. Also waiting in one of the lounges was Jemima Khan, wife of the former Pakistani cricket captain, Imran Khan. With Mrs Khan were her two young children, her brother and her mother, Lady Annabel Goldsmith, widow of the businessman Sir James Goldsmith.

That evening's BA 2069 service to Nairobi, callsign Speedbird 2069, was being operated by a Boeing 747-400 aircraft, registration G-BNLM. One of the last 747s to depart the North Terminal that day, the aircraft was parked nose-in to the building at Gate 54 as the crew prepared for the waiting passengers. On board the aircraft, the commander, Captain Bill Hagan and his co-pilot, Senior First Officer (SFO) Phil Watson, were proceeding through their departure checks. A third pilot, SFO Richard Webb, was also on board for in-flight relief purposes. The journey time from London to Nairobi, including taxiing times at either end, was eight and a half hours and, with the addition of pre- and post-flight procedures, produced a total duty time of ten hours. Although the 747-400 could be flown by only two pilots, the limitations on duty flight times did not permit two pilots to operate for ten hours on such a late evening departure. SFO Webb, therefore, was on board to take the place of each pilot in turn as he rested in a bunk to the rear of the flight deck, thus maintaining two pilots in the cockpit for most of the flight. All emergency procedures in the cruise, however, can be performed by only one pilot, so one of the two could take a short break for any quick visits off the flight deck that might be required. Such a process avoided the resting pilot being disturbed and the procedure worked well. Captain Hagan was taking his wife and two children with him on the trip, and they also waited in the terminal. The crew's stopover in Nairobi spanned New Year's Eve and the family had planned to be together for the celebration.

One feature of a late evening departure is that passengers are usually less stressed in the not so busy atmosphere of the terminal. The travellers often arrive early at the airport and are usually less active as they get tired. Many are also in transit from other flights. At check-in people are mostly quiet and at the gate they wait patiently for boarding. One passenger at the check-in desk for BA 2069 that evening, however, was not quite so stable and was drawing attention to himself. The gentleman in question was a Mr Paul Kefa Mukonyi, a man of rather large proportions, tall and well built, and with a shaved head. Mr Mukonyi was a Kenyan national returning home and at only twenty-seven was fit and strong. The Kenyan was simply dressed in a brown nylon jacket with light-coloured pants so, apart from his size, there was nothing unusual about his appearance. Something, however, was bothering Paul Mukonyi at check-in and he seemed confused.

Earlier that day Mr Mukonyi, who had been studying in France, had purchased a return ticket for his journey at a travel agent in Lyon and had paid almost £1,500 in cash. He had then travelled on British Airways from Lyon to London on an early evening flight and was transiting Gatwick Airport for the second leg of his journey home on BA 2069 to Nairobi. On such flights passengers and their baggage are normally checked-in at departure all the way through the transit stop to destination without further formalities. Seats are normally allocated for both sectors and boarding cards issued. Perhaps it was the fact that Paul Mukonyi had probably already been checked-in for BA 2069 in Lyon that was confusing the Kenyan, but there was no doubt that something was bothering him. At the check-in desk the staff did not view the gentleman as disruptive or violent, but were sufficiently concerned by his confusion to require him to be escorted to the departure gate. On arrival at Gate 54 under BA ground staff escort, the dispatcher, who was respon-sible for co-ordinating the departure arrangements, was summoned to talk to Mr Mukonyi. The Kenyan asked the dispatcher to call the police and, since the gentleman's bewildered state was obvious, he complied with the request.

As other passengers waited in line to board at Gate 54, two police officers arrived to interview the Kenyan. The gentleman was clearly confused and mumbled about guns being smuggled on board. The police managed to piece together from his ramblings that Paul Mukonyi believed he had been followed from France and that he feared his pursuer. He warned the police that the person following him might try to board the aircraft with guns and drugs. The police reassured the Kenyan that security was sufficiently tight to prevent guns being boarded and that he should not be concerned in that respect. Although, in the words of the police, his manner was 'paranoid, agitated and confused', he gave no indi-cation of a disruptive or violent nature. The officers could also find no evidence to substantiate his story. After interviewing the Kenyan for about ten minutes, the Sussex police officers decided that the captain of flight BA 2069 should be informed of his behaviour, and they escorted the Kenyan to the aircraft. At the door they consulted with the Cabin Services Director (CSD), who was the chief stewardess aboard. The police offered her their opinion that, although they were concerned by the man's incoherent talk and his apparent paranoia, their impression was that his confused condition was symptomatic of a fear of flying. Although they were not able to resolve the reasons for his confusion, the officers did not express concern that the Kenyan had mental health problems or that he was violent by nature. In addition, the officers explained, Mr Mukonyi had not committed any offence and they could not prevent him boarding by arresting him. In the opinion of the police, the Kenyan was simply a nervous flyer and was 'confused but fit to travel'. It was their duty to inform the captain of the man's behaviour, however, and they asked the CSD to liaise with Captain Hagan. After consulting with the captain on the flight deck, the CSD returned to inform the police that under the circumstances Captain Hagan would accept the passenger for travel. The police handed over the confused gentleman to the CSD but, before departing, advised that the crew keep an eye on the Kenyan and that his behaviour be monitored. Paul Mukonyi boarded the aircraft and made himself as comfortable as a very big man could in his 'tourist' seat in the World Traveller economy section.

As the last of the passengers took their seats, the doors were shut and the aircraft was made ready to depart. The 379 passengers, sixteen cabin crew and three flight crew, made a total of 398 souls on board. A few minutes after scheduled departure time, the 747-400 pushed back from the gate, taxied out to Runway 24 and took off into the fresh westerly breeze shortly before 23:00. The big jet banked after take-off, following the instrument departure, and turned to head east toward Dover, the Channel and the Continent. As Speedbird 2069 climbed over Northern Europe, the cabin crew began the drinks round. Mr Mukonyi, meanwhile, sat quietly in his seat and did not speak to anyone. He kept himself to himself, ignored his fellow passengers and bothered no one. It seemed that the police assessment of the situation had been correct and that the Kenyan had calmed down as the flight progressed. Unknown to anyone on board at the time, however, the big man's mind was anything but settled.

Paul Mukonyi was, in fact, a mentally ill and very sick young man who was suffering from paranoid delusions. A doctor might have described his condition at the time as a 'delusionary disorder of the persecutory type' and, although his persecution was imaginary, his fear was real. The Kenyan was studying tourism at university in Lyon and recently had been under a lot of pressure with exams. He was known to be a very bright student, but his present paranoid and anxious condition may have been a result of exam stress. No matter how his sickness had been triggered, however, there was no doubt in Mr Mukonyi's mind that he considered himself to be in mortal danger. While studying in France he had been overwhelmed with the fear that his life was at risk and he imagined he was being followed by people who were trying to kill him. He felt compelled to escape from Lyon and to return home for a short while to tell his parents that he believed his life to be in danger. The purchase of the return ticket on the spur of the moment from the travel agent earlier that day, and the payment of the ticket in cash, were evidence of his compulsion to flee. Mr Mukonyi's swift departure from Lyon, however, had not allayed his fears for, as he had mumbled rather incoherently to the police at Gatwick, he imagined his pursuers had followed him from France and that he was still at risk.

The Kenyan sat silently in his seat, but the remoteness from his fellow passengers was deceptive as he still feared his pursuers could be aboard. His dread that he was still at risk of attack may have been beginning to consume his mind but, if he was suddenly overwhelmed by fear, how would he act? The nature of the young man's sickness was such that his actions would probably not be planned or predetermined, and that any outburst resulting from his condition was likely to be spontaneous. As previous events had hinted, the big man was likely to be suffering terrible pressures resulting from his mind being crowded with confusion, anxiety and fear. For how long could Paul Mukonyi withstand the pressure? As the 747-400 approached its cruising altitude, the Kenyan sat like a ticking bomb that might explode at any time.

G-BNLM settled in the cruise at 33,000 feet (10,000 metres) and headed south-east bound across Europe, flying over France and Italy towards the southern Greek island of Crete and the eastern Mediterranean Sea. The meal service followed the drinks round and, as the trays were collected in, the passengers began to settle down for the night ahead. The 747-400 crossed the north coast of Egypt, just over eighty miles to the west of Alexandria, about three and a half hours into the flight, with about four and half hours journey time of the route over Africa remaining. In the absence of the moon the night was black, but clear, and the lights along the coast, plus the flares from oil wells, were plainly visible from the darkened flight deck. Speedbird 2069 proceeded southbound under Cairo Air Traffic Control. In the cabin the lights had been dimmed after the sale of duty-free, which followed the meal service, and the only illumination was from the glow of the nightlights and the flicker of movie screens. London time was now just before 02:30 on 29 December 2000. While the few passengers still awake watched films, the others, by now very tired, wrapped themselves in blankets in the hope of catching some sleep. Kenyan time was 05:30, three hours ahead of London, and the local dawn was expected in just over two hours. Estimated arrival in Nairobi was 10:00 local time.

Speedbird 2069 continued southbound, more or less following the course of the Nile as it meandered below. Not far to the east of the route lay the sights of man's ancient history: the pyramids at Giza, the Valley of the Kings at Luxor and the ancient temples of Rameses II at Abu Simbel, moved by engineers to higher ground to escape the flood waters of the Aswan dam. Following the route by day is an interesting exercise but in the blackness of the night the flickering of lights was all that could be seen. The weight of G-BNLM was now sufficiently reduced by fuel consumption to permit cruise at a higher level, and the aircraft was cleared to climb to 37,000 feet (11,300 metres). In the cabin, the flight attendants began their rest breaks, and off-duty crew members attempted to snatch some sleep in the cabin crew rest area situated above the ceiling at the tail of the aircraft. The remaining crew members manned the galleys or patrolled the quiet and darkened cabins. On the flight deck, progress continued as planned as the 747-400 crossed the border from Egypt into northern Sudan on airway Upper Amber (UA) 727. Speedbird 2069 was now no longer under radar control and was also outside the range of VHF (very high frequency) radio cover with Cairo. Reports would now have to be transmitted over long-range HF (high frequency) radio with Khartoum. HF radio is affected by atmospheric conditions and is frequently subject to static noise. On occasions it is even impossible to establish contact with a distant station. As a safety precaution all aircraft also transmitted position reports on 126.9 MHz, a frequency reserved for air-to-air communications over Africa. Meanwhile, the flight crew had also begun their rest breaks and Captain Hagan had retired to the crew bunk for some sleep. The relief pilot, SFO Webb, now occupied the captain's seat, and the two experienced first officers operated the radios, kept the logs up to date, maintained a weather eye on the radar screens for storm clouds and monitored navigation, engine and flight displays. The flight crew bunk area was to the aft and left of the cockpit and was very close by, so Captain Hagan could be summoned at a moment's notice if required. Behind the two pilots was the closed but unlocked door of the cockpit. Speedbird 2069 now fell within range of Khartoum VHF radio on 124.7 MHz and contact was established. Khartoum, lying at the confluence of the Blue and White Niles, displayed its lights ahead brightly in the blackness of the night, then slipped beneath the nose as the aircraft continued southbound on airway UA 727. One hundred and fifty miles to the south of Khartoum, with G-BNLM still cruising at 37,000 feet (11,300 metres), the route lay abeam the town of El Obeid in south-central Sudan. With progress so quiet, SFO Richard Webb took the opportunity to pop out of the flight deck to use the toilet just aft of the flight deck door, leaving his colleague, SFO Phil Watson, alone in the right-hand seat. The time now was 04:45 in the morning London time, 07:45 Nairobi local time. The sun still lay just below the horizon to the east, but the diffuse light of the almost new day was beginning to permeate the grey December morning. Behind the flight deck, in the upper deck section of Club Class, the passengers were asleep or half asleep, apart from two children watching a film. Likewise, in other sections of the aircraft, people were mostly asleep or dozing in the quiet and darkened cabins. Only part of the cabin crew was on duty while the others rested and only one pilot remained in charge on the flight deck. If the fate of unfortunate timing was to play a part and anything untoward was to happen now, the moment could not have been worse. In the meantime, a large Kenyan man in the World Traveller economy section had stirred from his position, stepped into the aisle and had begun moving towards the forward end of the aircraft.

What was happening in Paul Mukonyi's head as he wandered in the cabin was difficult to say, but he appeared to be suffering some form of paranoiac panic attack. It seemed his anxiety and fear had completely deranged his mind and that he had lost control of his actions. His movements, however, were more confused than purposeful. He did not march determinedly in a positive direction but ambled up the aisle in a hesitant and uncertain manner. The Kenyan did not appear to know where he was going and, as he strolled in a confused state in the cabin, he mumbled incoherently to himself. Had the big man proceeded in a rushed and aggressive manner his actions might have caused alarm, but as he ambled quietly and mumbled softly to himself his movements went mostly unobserved. Mr Mukonyi continued his stroll through the World Traveller cabin and, after a while, found himself at the bottom of the stairs leading to the part of the Club Class section on the upper deck. Slowly he began to mount the staircase. The Kenyan ambled past the children watching a film in the upper deck Club Class section, past the sleeping Brian Ferry and his family, and approached the unlocked cockpit door. Outside the flight deck he hesitated, appearing, once again, to be confused as to his actions and intentions. For a short while he hovered in the vicinity of the flight deck, mumbling constantly, but whether talking to himself or praying was not apparent. After a few minutes Mr Mukonyi appeared to have resolved his confusion and he moved to open the cockpit door. Once again, had his actions been rushed and noisy, he would have drawn attention to himself, and he might even have alerted SFO Webb in the nearby toilet, but he continued to move silently. Slowly he turned the door handle, entered the cockpit quietly and gently closed the flight deck door behind him. To SFO Watson, sitting alone at the controls in the right hand seat, Mukonyi's entrance was no different than if his colleague, SFO Webb, had returned to the flight deck.

Without warning and without even uttering a sound Mukonyi threw himself across the cockpit at SFO Watson. Whether the Kenyan imagined he could protect himself by seizing control of the aircraft or end his predicament by committing suicide is not known, but his drastic action had a devastating effect. SFO Watson turned sharply at the movement to find diving towards him a huge hulk of a man who was so bulky he seemed to the first officer to be wearing at least six layers of clothing. The first officer noticed a determined look of intent in Paul Mukonyi's eyes as he thrust himself over the central pedestal that lay between the pilots' seats. The Kenyan lunged over Watson's left shoulder and leaned right over him, grabbing the control column and pulling violently. The autopilot

13

was engaged, as is normal in the cruise and, when required, is usually disconnected by the pressing of a button on the control column. The force of the pull was so severe, however, that the autopilot disconnected itself. SFO Watson hit the big Kenyan and tried to put him in an arm lock. He also shouted loudly for help. The nose pitched up 26° above the horizon and the aircraft began to climb, with the wings rocking from side to side as the deranged man wrestled with the first officer and the controls. The violent inputs to the control column were such that the bank angle oscillated 30° to the left and right. Phil Watson's view of the flight instruments was totally obscured by the upper torso of the Kenyan as he lay across Watson's lap and pinned him to his seat. He could feel the wild movements of the aircraft but without a view of the instrument displays the first officer was unaware of the flight profile. 'He was all over the place,' stated SFO Watson, 'leaning across the instrument panel and between me and the windscreen so I couldn't see anything outside.' As the speed decayed, the autothrottle, which had remained engaged although the autopilot had disconnected, sensed the loss of speed and advanced the thrust levers to apply full power. Phil Watson fought with all his might to free Mr Mukonyi from the controls, but his actions were in vain. Captain Bill Hagan, in the crew rest area to the rear of the flight deck, was immediately roused from his slumbers by the noise and commotion and jumped up from the bunk. Quickly he leapt into the cockpit and raced to assist his struggling colleague. 'It was obviously pretty frightening', the captain stated. 'When I got back on the flight deck there was this guy built like Mike Tyson fighting with the first officer, who was flying the aircraft. He was like a man possessed.' Captain Hagan immediately jumped on the attacker's back and tried to wrestle him away.

The aircraft's pitch-up attitude, however, was maintained by the big man still pulling back on the control column and the indicated airspeed continued to reduce. When flying at 37,000 feet (11,300 metres) the 747 cruised at Mach 0.845 with an indicated speed in the rarefied atmosphere of 280 knots on the airspeed tape. G-BNLM climbed 3,000 feet (900 metres), ascending to 40,000 feet (12,200 metres), and the speed loss became critical. The indicated airspeed, despite the autothrottles still applying full power, now fell dangerously low to 165 knots. At 40,000 feet (12,200 metres), with an indicated airspeed of only 165 knots, there was insufficient airflow over the wings to sustain the 747 in flight, and the aircraft stalled and dropped from the sky. The nose plunged from an attitude of 26° above the horizon to 20° below, while the left wing dropped sharply as the aircraft banked steeply to the left. As the 747 plummeted earthwards out of control Hagan and Watson fought desperately to free the strong Kenyan from the control column.

In the cabin, the violent movement of the aircraft caused extreme alarm amongst the mostly drowsy passengers in their seats. To make matters worse, the struggle on the flight deck appeared to have caused some switches on the pilots' overhead to be knocked. The night lighting extinguished in the cabin, leaving the passengers in darkness and adding to the confusion. The sensation of dropping from the sky was terrifying as the aircraft nose lurched downwards, and some of those resting without seat belts fastened were flung around in the blackness. The emergency oxygen switch also appeared to have been triggered by the struggle and oxygen masks dropped from the ceiling, further alarming the occupants. Hand luggage tumbled from overhead lockers that had flown open and other items on seats were thrown around. Many people screamed with fright. Jemima Khan's brother, Benjamin Goldsmith, stated that 'suddenly the plane went into this violent, violent dive, like shuddering, and went very, very steeply downwards. Everyone was woken up by the screaming.' Katie Laybourne, a nineteen-year-old student going on holiday with a friend, said, 'We dropped and dropped. It was a real nightmare.' A doctor from Portland, Oregon, also described the scene. 'A couple of people bounced up quite a bit, the plane was making some very weird noises. It was like a roller coaster.'

On the flight deck the situation was getting out of hand as the two pilots still failed to dislodge Mr Mukonyi from the controls. In less than a minute from the launch of the attack by the big Kenyan, all 398 people on board were now in mortal danger as the 747 plunged from the sky. The left wing continued to drop and the wings rolled through the vertical to 94°. The nose still pitched down 20° below the horizon and thousands of feet had been lost. If the aircraft were to continue its roll manoeuvre it might roll on its back and recovery might be impossible. Desperate circumstances require desperate measures and Captain Hagan felt that the only thing to do to break the big man off was to stick his finger in his eye. Hagan poked his finger hard into the attacker's right eye and Mukonyi winced in pain. The action had the desired effect for, to protect himself, the Kenyan finally slackened his iron grip on the control column. With the big man in so much pain Bill Hagan thought he had a chance of pulling him away and with all his strength he began to drag the Kenyan off. SFO Richard Webb, who had left the flight deck only a few minutes earlier, rushed back to the cockpit at the sound of the commotion and assisted his captain in removing Mukonyi. In Club Class, close behind the flight deck, sat Mr Clarke Bynum, a tall ex-basketball player from South Carolina. He had been awakened by the shouts for help from the flight deck and the dramatic changes of engine noise. From his seat he could see by the horizon visible throughout the window that the aircraft was in a dive. He felt compelled to intervene and, as he opened the cockpit door, saw what appeared to be Mukonyi wrestling with the first officer in the seat and the other two pilots trying to drag him off. Hagan and Webb, using all their strength, finally managed to pull the deranged man clear and, as they did so, the tall Mr Bynum was able to assist in restraining him. Unfortunately, as the Kenyan released his grip on the controls, the pulling back of the column induced nose-up inputs which tightened the turn and made a bad situation worse. SFO Phil Watson, still strapped in the right-hand seat, was now able to view the flight displays and get his hands back on the controls. It was going to be a difficult task, however, to get the aircraft under control after such violent manoeuvres and to recover from such an unusual attitude. By now the 747 had dropped about 8,000 feet (2,400 metres) from its peak of 40,000 feet (12,200 metres) and was still descending rapidly in a steep dive with the wings banked through the vertical and the aircraft threatening to roll on its back. Fortunately, however, the autothrottles had remained engaged throughout the struggle and had automatically closed to idle power as the speed had increased sharply in the steep descent. Fortunately, also, SFO Phil Watson was a cool and able pilot.

All pilots, of course, undergo training in recovery from unusual or extreme attitudes, but the circumstances can be very dramatic and few have any desire to experience such events in real life. Watson was aware that the use of rudder on a big jet to control excessive roll at high speed could be ineffective and could cause structural problems. He was also aware that at a large bank angle the nose of the aircraft drops into the turn and that the use of elevators at a large nose-down attitude only exacerbates the effect. SFO Watson knew the drills and, under extreme pressure, with the flight in imminent danger, applied them skilfully. With the 747 still plummeting from the sky, he calmly and smoothly rolled the wings level using only the ailerons and by carefully observing the flight displays. With the gathering light the horizon was also now visible in the sky and assisted his efforts. Once the wings were level the next task was to arrest the descent and pull the aircraft from the dive. If the first officer pulled back too firmly the aircraft might suffer structural problems and if too slowly the 747 might lose excessive altitude. Although the 747 was still about 32,000 feet (9,700 metres) above the ground, there were many aircraft flying north and south on the airway at different levels and there was a risk of collision. There was also a risk of being unable to rescue the 747 from the dive if the speed became excessive. Steadily, smoothly and firmly SFO Watson began to pull back on the control column.

Mukonyi had now been dragged from the controls but he was still on the flight deck and still a danger. Captain Hagan was concerned that Mr Mukonyi might attack again and he fought desperately to get the man off the flight deck. In the mêlée Hagan had his ear bitten and, as he pushed the Kenyan's face, the big man bit into the captain's finger and almost severed it. As Hagan and Webb struggled to eject Mukonyi by pushing him backwards, Bynum managed to get the Kenyan in an arm lock round his neck and to force him down. Finally they were able to drag the intruder from the cockpit and to subdue him by pinning him to the floor.

As the ferocious fight continued, SFO Phil Watson still struggled with the controls to rescue the situation. The first officer continued to pull back steadily on the control column to ease the 747 from its dive. Watson pulled 2.3 g as he arrested the descent but managed to maintain the indicated airspeed below 240 knots. The first officer observed 30,000 feet (9,000 metres) on the altimeter tape as the 747 bottomed out of the dive and the aircraft then climbed another 1,000 feet (300 metres) before he finally regained full control. The 747 had plunged 10,000 feet (3,000 metres) in thirty seconds and had been recovered from an extreme attitude with the aircraft diving and almost rolling on its back. Phil Watson had done a magnificent job. Captain Hagan now had the chance to assist his first officer and rushed back to his seat. 'By this stage,' declared the captain, 'there were red lights flashing all over the cockpit.' As Watson levelled G-BNLM at 31,000 feet (9,500 metres), Hagan immediately called on the air-to-air frequency of 126.9 MHz to warn other traffic that they had lost altitude and were no longer cruising at their cleared level of 37,000 feet (11,300 metres) nor heading southbound along the airway. In the cabin the passengers were alarmed and frightened by the violent manoeuvring of the aircraft and the dramatic changes in engine noise. Many were very upset and there was a lot of crying.

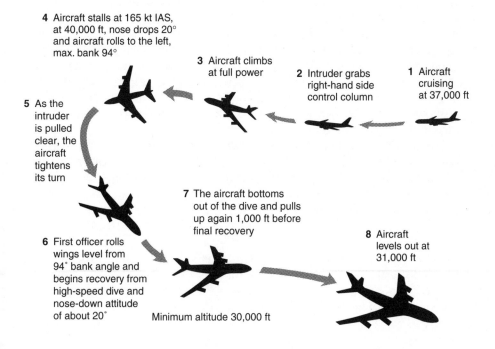

4 Aircraft stalls at 165 kt IAS, at 40,000 ft, nose drops 20° and aircraft rolls to the left, max. bank 94°

3 Aircraft climbs at full power

2 Intruder grabs right-hand side control column

1 Aircraft cruising at 37,000 ft

5 As the intruder is pulled clear, the aircraft tightens its turn

7 The aircraft bottoms out of the dive and pulls up again 1,000 ft before final recovery

8 Aircraft levels out at 31,000 ft

6 First officer rolls wings level from 94° bank angle and begins recovery from high-speed dive and nose-down attitude of about 20°

Minimum altitude 30,000 ft

Amazingly, however, no one had been seriously hurt and only four passengers and a stewardess received minor injuries. Captain Hagan, still out of breath from his struggle and barely able to talk, quickly spoke on the PA to reassure his passengers. 'A nasty man just tried to kill us all,' he explained, and then added in reassurance, 'but we are all right now.' Captain Hagan then went on to say that the aircraft was sound. The entire episode, from the intrusion of the Kenyan into the flight deck to the recovery of the aircraft from disaster, had taken barely two minutes.

Outside the flight deck, Mukonyi was now firmly pinned to the floor by a further three passengers who had joined in to help and who were sitting on top of him. The Kenyan kept mumbling to himself making little or no sense. By now other crew members had arrived with restraining devices and they handcuffed Mukonyi. They also brought with them extension seat belts that they used to tie round his arms and ankles. As BA 2069 resumed its journey to Nairobi, normality returned to the flight deck. In the upper deck cabin Mr Mukonyi was lifted bodily to the rear and tied to a seat for the remainder of the trip. Later Captain Hagan left the flight deck to visit the cabin to check on the welfare of his passengers and to make sure that his family were all right.

About two and a half hours later, at 10:10 local time, BA2069 landed safely at Nairobi, much to the relief of all on board. The police were on hand when G-BNLM arrived at the gate and they immediately arrested Mr Mukonyi. He was heavily sedated and taken to hospital, where he was guarded by armed officers. Later, a doctor at a psychiatric hospital in Nairobi stated that Mr Mukonyi should not stand trial. 'He is a patient,' explained the psychiatrist, 'and he is in the hands of a doctor. That is where he belongs.' BA 2069 was due to continue on the next leg of its journey to Dar es Salaam in Tanzania and, after engineering checks, the aircraft was permitted to proceed on its way. The incident had been a close-run event and, but for the skill, quick thinking and courage of the crew, could have ended in tragedy.

In May 2001, Captain Bill Hagan, Senior First Officer Phil Watson and Senior First Officer Richard Webb were presented with the Polaris Award by the International Federation of Airline Pilots for their outstanding efforts. Six months later they also received the British Airline Pilots' Association Gold Medal for bravery.

Chapter 2

That Falling Feeling

It has been calculated, for the benefit of free-fall parachutists, of course, that it takes thirteen to fourteen seconds for the human body to reach 99 per cent of its low-level terminal velocity, a speed of almost 125 mph (200 km/hr), after falling just over 1,800 feet (550 metres). The figures assume normal atmospheric conditions and that the person is dropping in a random, tumbling manner. If the sky diver plummets earthwards in a head-down position, speeds of 185 mph (300 km/hr) are achieved after the same time and descent.

It has not been unknown in the past, however, for unfortunate individuals to have tumbled without parachutes from balloons, airships and aircraft, and the stories are legion of those who have survived to tell the tale. As early as 1839, a Mrs Graham fell 100 feet (30 metres) from a balloon over Britain and landed on soft grass with little injury. She was wearing a long ankle-length dress which billowed out as she sank, effectively braking her fall, and she touched down firmly with her life, if not her modesty, intact.

In the mid-1930s, during the golden era of American airship operation, one young man came closer to falling to certain death than can be imagined. During the early years of Franklin D. Roosevelt's presidency, he undertook a voyage, courtesy of the US Navy, by cruiser from Washington DC to Hawaii, via the Panama Canal. While sailing across the mid-Pacific en route to Honolulu, a US Navy Airship, the *Akron*, launched in August 1931, was dispatched from San Francisco to drop the latest daily newspapers on the deck of the president's ship.

The flight entailed a stop at a military camp near San Diego in Southern California, and sailors from the nearby naval base of North Island were commandeered to act as ground crew. The *Akron*'s arrival date was announced at short notice and preparations were somewhat hasty. No one in the area had seen such a large airship before and training was rather primitive and poor. Instruction centred around the guy ropes, which at their lower end had a multitude of lengths and handles, like a giant cat-o'-nine-tails. The sailors were broken into teams, one group per guy rope, and each team member was told to grasp one of the many handles and 'to hold on for dear life'.

As the *Akron* approached the camp to moor, the main wire, which held the airship in place, was secured and the guy ropes were lowered. Team members each grasped a handle and pulled down on the ropes to manoeuvre the giant ship into position. As the sailors laboured at their task, a gust of wind suddenly caught the airship, the main wire parted and the nose rose into the air. The captain, Scott Peck, yelled from a microphone, 'Let go, let go, let go'. Most sailors, on hearing the command, were happy to be free of their charge, in spite of earlier instructions to hold on. One young able seaman, however, unwilling to drop on his colleagues below, was whisked into the air and in moments found himself hundreds of feet above the crowd, getting higher and higher by the second. The people below gasped in horror as they saw the young recruit dangling high in the sky, holding on to a flimsy handle with all his might. Keeping a cool head, and being aware he couldn't hold on for long, the sailor realised his only chance of survival was to secure himself to the rope. By an amazing feat of bravery and strength, he pulled himself up with one hand and at the same time managed to wrap the remainder of the rope round his body and to secure it firmly at his waist. He was thus able to let the rope take the strain of his weight. As the *Akron* climbed, the young man simply dangled below until, about 3,000 feet above the crowd, the airship crew managed to winch up the guy rope and pull the sailor aboard to safety.

In a slightly more recent incident, in 1972, a young woman fell from an aircraft several miles high in the sky, but by some miracle managed to survive the plunge. The event is noted in history as being the longest fall without parachute and is recorded for posterity in the Guinness Book of Records.

The aircraft, a DC-9, registration YU-AHT, of Yugoslav Airlines (JAT) took off from Copenhagen at 15:20 GMT on 26 January 1972, and climbed southbound over the then German Democratic Republic (GDR) en route to Yugoslavia. The DC-9 flew east of Berlin, following the Upper Amber Four (UA4) airway, and levelled off at 33,000 feet (10,050 metres) under the control of Cottbus Air Traffic Control Centre (ATCC) on 126.7 MHz. On board, the passengers settled down for the journey ahead, while the cabin crew, one of whom was a young stewardess named Vesna Vulovic, busied themselves serving refreshments. Not in her wildest dreams could Miss Vulovic have imagined what was going to happen to her next.

At just after 16:00 GMT the DC-9 passed over Hermsdorf radio beacon on the border with Czechoslovakia and was instructed to call Prague ATCC on 132.8 MHz. The Yugoslav aircraft's flight routeing was then planned to continue along UA4 to overhead Prague, along UA15 to Vienna in Austria, and on to Yugoslavia. Shortly after the instruction to change frequencies, while flying over the Czechoslovak town of Serbska Kamenice, an explosive device detonated and Vesna Vulovic found herself blown from the aircraft. At 33,000 feet (10,050 metres), the outside air temperature (OAT) over Europe in January is around –60°C, and the young stewardess was instantly unconscious, tumbling earthwards like a rag doll. She dropped the first 22,000 feet (4.2 miles/6.7 km) in two minutes and by the last 11,000 feet (2.1 miles/3.3 km) had reached her low-level terminal velocity of 125 mph (200 km/hr). Instant death lay one minute below. As Miss Vulovic reached the end of her high-speed fall, the first thing she struck was a thickly wooded bank of tall pine trees. She crashed through the trees, breaking twigs and snapping branches as she plummeted towards the ground. Each impact with branch or bough, however, broke her fall and reduced her velocity so that by the time she hit the earth her speed had been slowed considerably. She finally struck the soft forest floor three minutes after being ejected from the DC-9, unconscious and badly injured, but very much alive. It was a quite amazing occurrence. Vesna Vulovic spent months in hospital recuperating from her injuries but went on to make a full recovery.

As recently as 1990, another incident occurred that was to capture the attention of the world. The event, involving a British Airways (BA) British Aircraft Corporation (BAC) 1-11 aircraft, demanded courage, strength and resourcefulness and put those involved to a severe test. The incident happened on a quiet Sunday morning on 10 June 1990, but the tale began to unfold some one and a half days earlier on the night of 8/9 June.

On the evening of Friday 8 June, the weather in Birmingham, England, a city lying 100 miles (160 km) north-west of London, was partly cloudy with the possibility of rain showers developing throughout the night. The BAC 1-11 aircraft in question, registration G-BJRT, was a 500 series powered by two Rolls-Royce Spey engines. The airliner was due to undergo some routine maintenance that night and had been pushed back into a hangar on the eastern apron at Birmingham International Airport with its nose pointing outwards through the open doors. The hangar was only just large enough to contain the aircraft and it was a tight fit. Close by the hangar were associated workshops and a manned store, while at the far side of the airport, beneath the international pier, were other small workshop areas and an unmanned store room. A carousel there comprising over 400 drawers contained consumable spares, such as nuts and bolts, which were available on a self-serve basis.

The shift maintenance manager (SMM) responsible for the night's work routine arrived forty-five minutes early in order to catch up with paperwork and to organise the shift which began at 21:30 GMT, 22:30 local time (local time used throughout). It had been a while since he had worked during the hours of darkness, as he had been on holiday during the last night shift and this was going to be his first night out of bed in five weeks. Earlier in the evening he had managed to sleep only about one and a half hours, which was hardly sufficient, so by the small hours of the next morning he would possibly be feeling the strain.

Almost all maintenance activity was conducted at night for operational reasons as most aircraft flew a full programme during weekdays, commencing at 06:30, with a reduced service at weekends. The men of the night shift, therefore, usually undertook the necessary airworthiness tasks. The SMM had an exemplary record with British Airways (BA), including receiving commendations, and his crew were an enthusiastic team who were proud of their achievements in successfully completing assignments.

With the manager on duty that night were four other engineers who would assist with the work schedule, the complement being less than week days because of the quieter weekends. The team, however, still had to fulfil the normal maintenance and rectification requirements as well as offer flight servicing and refuelling to other contracting airlines which operated in and out of Birmingham during the night. Amongst the maintenance tasks to be completed that night shift were a number of cabin faults, some routine items and three significant defects, one of which was the captain's windscreen on G-BJRT, which needed to be changed. On a previous flight the captain had raised an air safety report indicating that, during the cruise, darkening and bubbling had been noticed in a small area at the bottom of the left-hand flight deck windscreen. The window was obviously leaking air from the pressurised interior and its replacement was necessary.

Throughout the night, as the engineers worked at their tasks, the SMM divided his duties between administration, engineering and flight servicing. After his meal break, during which he continued with his office duties as he ate his sandwiches, he directed his attention to G-BJRT. The BAC 1-11 was not due out on service until Sunday morning, so there was no pressure on that account, but there was a problem regarding its maintenance schedule. At 06:30 on Saturday a team had been booked, using overtime, to wash the aircraft. The previous week a similar team had been booked and not used, wasting money, and the SMM was anxious not to repeat the error. Since the BAC 1-11's window would have to be changed before the wash, the task would have to be accomplished during the night's work schedule. The manager had changed windscreens before but, to refresh his memory of the procedure, he briefly read the maintenance manual and satisfied himself that it was a straightforward job.

Some tasks, however, are not quite as straightforward as they seem, and the window change was to be no exception. Also, as in almost all incidents, it is usually a chain of events which leads to a happening rather than a single error, and this story was to follow a similar pattern.

It could be argued that the affair began at least thirteen years before at the time of manufacture of the BAC 1-11 in 1977, or perhaps even earlier, with the design of the aircraft. Design philosophy, then, was of another age, and the pilots' windscreens, for example, were arranged to be bolted in place from the outside, rather than like a plug from the inside, which would have been safer. With pressurisation of the aircraft's hull, a plug-fit windscreen would seal more firmly while, with the actual design, the bolts took more and more strain as the differential pressure between inside and outside the aircraft increased. Any loose or improperly fitted bolts would degrade the situation and it was imperative that the correct bolts were used. A new windscreen, therefore, would have been better supplied in kit form with the right type of bolts included, but this was not the case. In fact, during a window replacement, it was accepted practice that new bolts were only used to replace old ones where the latter were damaged or worn. It was up to the engineer completing the job to obtain the correct bolts from the stores, a task not quite as simple as it might seem. A number of the bolts used on the BAC 1-11 were British Standard A211, with suffixes appended to classify individual bolt types, e.g. A211-8D, A211-7D, A211-8C, etc. The bolts were supplied in clearly marked packets but could not be individually stamped with the type identifier, e.g. 7D, because of their small size so, once opened, had to be placed in properly labelled drawers. If necessary, loose bolts could, with care, be identified by comparison, but it was not an easy task to achieve.

The BAC 1-11 pilots' windscreens used ninety 8D bolts, which were surprisingly small for the task in hand, having a shank length of only $8/10$ in (20 mm) and a diameter of $3/16$ in (4.15 mm) and were fitted from the outside to hold the window in place. The bolts were fed through holes in the windscreen's rim and screwed into anchor nuts, which were permanently fitted to the window frame. The 8D-bolt specification was defined in the BAC 1-11 500 series parts catalogue which, on the relevant page, also displayed a sketch of the pilots' windscreens. On the same page were drawings of the adjacent direct vision (DV) side window, which could be opened on the ground by sliding the window inwards and backwards. The DV window required 7D bolts, the same diameter as the windscreens' 8D bolts, but a little shorter. It would have seemed a fairly simple design task to have used the same size of bolt for both window types, but this had not been done. In the parts catalogue, however, only one bolt was illustrated beside the window sketches, the 7D bolt, and this had led to confusion. Indeed, a number of BAC 1-11s, in both BA and other airlines, had had their pilots' windscreens incorrectly fitted with mostly the shorter 7D bolts. G-BJRT was included in that group, although the windscreen had been fitted four years earlier before the aircraft had been purchased by BA. The 7D bolts, therefore, although the wrong length, were clearly adequate. The recent leaking of the windscreen, however, may have been a final indication of the unsuitability of the 7D bolts.

The SMM began to remove the captain's defective windscreen at 04:00 with the help of one of the other engineers. The window was heavy, being 60 lb (27 kg) in weight, and consisted of five-ply glass/vinyl construction. It was heated to improve its impact resistance at low temperatures in the event of a bird strike and to prevent misting. An inspection of the old seal found it to be serviceable so the new windscreen was simply manoeuvred into position. Apart from help in lifting the windows in and out, plus some minor assistance, the SMM worked on his own while the other engineers continued with their duties.

On removal of the bolts from the windscreen, a number were found to be defective and required to be replaced. Rather than inspect the parts catalogue, which was on a microfiche reader, for the correct bolt specification, a time-consuming exercise, the manager took one of the removed bolts to the manned store to cross-check the size. He correctly identified by comparison the bolts as 7Ds and, unaware that these were the wrong type, asked the storeman for a fresh supply. The storeman stated that the longer 8Ds were used to fit that windscreen but did not press the point. The SMM was aware that older aircraft can have differing modification states with, perhaps, 7D standard window bolts on one and 8Ds standard on another, so he was not convinced by the storeman's comment. The 7Ds had been removed from G-BJRT and had obviously proved to be satisfactory so, unaware of his error, 7Ds he would use. They had survived four years with the old windscreen and would last just as long with the new. Now, however, another problem arose; when the storeman checked for spares he found that there were virtually no 7Ds left. The only other source of bolts was in the carousel at the far side of the airport, so the SMM jumped in his car and drove through the darkness to the workshops below the international pier. Rain showers had been passing through the night and the tarmac was still wet.

The time was now about 05:00 and, in the small hours of the morning, the manager was at his lowest ebb. It being his first night out of bed in five weeks he was probably suffering from sleep deprivation. At the spares carousel he attempted once again to acquire the bolts he needed by matching the 7D bolt from the old windscreen with those in the drawers. This time, however, the circumstances were different. He was now on his own, the area was not well lit, and he did not have his reading glasses with him. Although the spectacles were only mildly corrective, they might have helped under the circumstances. After what he assumed to be a careful comparison, the SMM selected, in error, 8C bolts instead of 7Ds. The length of the 8C shank was the same as the correct 8D shank but, because of a smaller head size, the 8C's overall length was almost exactly the same as the 7D's overall length. The main problem with the 8C bolt, however, was that its diameter was fractionally less than the 7D and 8D bolts. If used, the 8C bolt could be dangerous, but the saving of the situation would be the inability of the 'C' category diameter bolt to engage with the threads of the anchor nuts which were designed to accept the 'D' category diameter bolt – or would it?

The manager returned to the hangar and, using six 7D bolts from the old windscreen and eighty-four new 8C bolts from the carousel, began to secure the window to the frame. The hangar doors were closed because of the rain showers outside and, with only a few feet (less than one metre) between the aircraft nose and the doors, the work area was cramped. It was impossible to place the raised platform on which he was standing in the correct position along the side of the aircraft because of the narrow space and, as a result, most of the bolts could only be reached by stretching outside the safety rail across the nose of the 1-11.

As each bolt was fed into place it seemed to screw down satisfactorily. By an unfortunate circumstance, the 8C bolts, although fractionally smaller in diameter, had the same thread pitch as the 'D' specification bolts and they appeared to engage adequately in the anchor nuts. When any force was applied, however, the threads would most certainly slip. In order to ensure each bolt was firmly secured, a torque-limiting screwdriver was employed to supply the correct pressure. When the manager screwed the bolts down tightly, the 8C bolt threads, as expected, began to slip in the anchor nuts but, by a further unfortunate circumstance, the movement of the threads slipping was similar to the action of the torque-limiting screwdriver clutch releasing when the correct torque level was reached. The SMM was, therefore, left with the impression that the bolts were properly in place and correctly tightened.

One other clue that was available was the smaller size of the 8C bolt heads. Each bolt hole was countersunk so that the bolt head lay flush with the windscreen rim and, when an 8C bolt was screwed in place, the smaller head could be noticed (see diagram). The inadequate access the manager had to the work area, however, made detection more difficult, and the indication went unobserved.

The manager finished replacing the new windscreen by about 06:00, which left adequate time for the 1-11 to be removed from the hangar and made ready for the wash team arriving at 06:30. As per the approved certification procedure, a task such as a window change did not require the

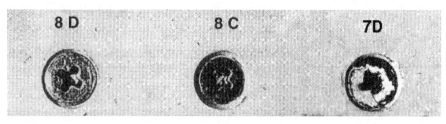

The heads of sizes 8D, 8C and 7D bolts.

work to be supervised or double-checked, and the job was signed off by the manager as having been completed. Had the work been conducted on a vital system, such as a flying control, for example, a duplicate inspection would have been necessary. There was also no requirement to carry out a pressurisation check on the ground, and the first time the windscreen would be put to the test in earnest would be when the 1-11 embarked on its first flight the following morning. On G-BJRT's initial climb, pressure on the windscreen would build. Would there be sufficient strength in the slipping threads to hold the window in place or would the bolts be stripped out of the anchor units? If the windscreen did blow, would the captain's seat belt hold him in place, or would the exhausting air force him into the atmosphere? And if he did tumble earthwards could he, like the Yugoslav flight attendant, survive the fall? Like a bomb waiting to be detonated, the 1-11 sat undisturbed on the tarmac throughout Saturday and Saturday night awaiting departure of the next day's flight.

Sunday 10 June dawned a pleasant day with some scattered cloud between twelve and fifteen thousand feet, a light northerly wind, visibility 10 km and a temperature of 15°C. G-BJRT was scheduled to operate flight BA5390, a regular service from Birmingham International Airport direct to Malaga, Spain, departing at 08:00. For the crew of BA5390, Captain Tim Lancaster, Senior First Officer (SFO) Alastair Atchison, Purser John Heward and flight attendants Nigel Ogden, Sue Prince and Simon Rogers it was an early start, for report time was at least one hour before departure. The captain was forty-two years of age and had amassed over 11,000 flying hours with just over 1,000 on the BAC 1-11, while the first officer, also a very experienced pilot, had 7,500 hours to his credit and 1,100 hours on the type. The two men had never met before, so introduced themselves to each other in operations before examining the paperwork. Being a nice summer's day, the trip to Spain was expected to be routine, except for the usual congestion over Europe, which was causing delays. The earliest air traffic control airborne slot time (take-off time) available was 08:20, so departure would be late.

It was agreed that the first officer would fly the aircraft that morning so, as the pilots conducted their pre-flight checks in the cockpit, Captain Lancaster discussed the technical log history with his co-pilot. Both noticed that the windscreen on the captain's side had been changed early the previous morning and since then the aircraft had not flown.

At just after 08:00, the engines were started and, with eighty-one passengers on board, mostly holiday-makers bound for resorts along the south coast of Spain, the aircraft taxied out to runway 33 in time for the pre-arranged departure slot. The 1-11 weighted 42.9 tonnes, 10 tonnes of which was fuel for the two-and-a-half-hour flight. BA5390 lifted off at 08:20 and soon after take-off turned southbound for Airway W5, which directed the flight towards the south coast of England, near the Isle of Wight. From there the 1-11 would proceed over the Channel, across France and on into Spain. The landing gear and flaps were raised in sequence and when settled in the climb the after-take-off checks were completed. The autopilot was engaged, the crew began to relax a little and shoulder harnesses were released and lap straps loosened. At this stage Captain Lancaster took over the flying of the aircraft, as per BA standard operating procedures, and would hand back control to his co-pilot for the landing into Malaga. He accelerated the twin-engined jet to its climb speed of 300 knots.

Air traffic control (ATC) first transferred from Birmingham Radar to the Daventry sector controller of London Air Traffic Control Centre (LATCC), then from there to the LATCC Bristol sector frequency of 132.8 MHz where the controller instructed the 1-11 to climb under radar control to flight level (FL) 140 (14,000 feet). The aircraft, callsign Speedbird 5390, now approached the large, busy London area from north-west of Heathrow Airport and was issued a series of headings to steer to remain clear of traffic. The captain adjusted the autopilot heading control to comply with the instructions and monitored the climb while the co-pilot monitored the radio. It was a beautiful day and, as the 1-11 flew over Oxfordshire, the pilots had a wonderful view of the Thames Valley below. They could see the great river meandering for miles through the centre of the county and they could pick out sights along their route. The town of Didcot lay ahead with Wallingford to the east and Wantage to the west. The two men on the flight deck talked about the countryside below, pointing out the various towns. To the crew's right just north of Wantage, was the village of East Hanney, the captain's home.

'I live just over there', he remarked.

Further instructions were received from London's Bristol sector controller.

'Speedbird 5390, turn right heading 195, climb FL230.'

SFO Atchison acknowledged the radio call while the captain turned the aircraft right and continued the climb. The time was now 08:30, and in the cabin the flight attendants were preparing to serve breakfast to the passengers. One of the stewards, Nigel Ogden, popped into the flight deck with hot drinks for the crew. A few moments later Speedbird 5390 passed overhead Didcot, climbing through 17,000 feet with an outside air temperature (OAT) of 17°C and the wind blowing from the north at 17 knots. Both pilots monitored the flight's progress and, as the co-pilot adjusted his compass heading cursor, the captain looked ahead through the window, checking the skies in front. Suddenly, passing 17,300 feet, at a climb speed of 300 knots, a movement caught Captain Lancaster's attention and he could scarcely believe his eyes. His windscreen began to move slowly out of its frame. He just sat there stunned, mind blank, unable to comprehend what he was seeing. The entire episode undoubtedly happened in micro-seconds but, to Tim Lancaster, the action was unfolding in slow motion. As if in a time warp, he watched the windscreen gradually ease out of its frame until, having risen about 6 in (15 cm), the scene exploded with a tremendous bang. The 60 lb (27 kg) window shot rearwards, dented the fuselage, ripped off the high frequency radio antenna and tumbled earthwards, finally coming to rest in a field near Cholsey, just south of Wallingford. With the instantaneous reduction in pressure the air in the cabin misted as the moisture condensed. At the same moment, the pressurised air trapped within the hull vented through the open windows and exhausted to the atmosphere in about 1¼ seconds with a force equivalent to 2.4 tonnes. Captain Lancaster was ripped from his lap strap, his head hit the roof and his torso was drawn through the window. The noise of the rushing air escaping was enormous, and loose objects such as flight logs, pens, charts and clip boards, etc., fluttered out into the sky. A steward's jacket was later found in Reading High Street, an ID near Newbury and the captain's headset in a farmer's field. The flight deck door tore from its hinges and blew into the cockpit, shattering against the centre pedestal between the two pilots. The door was designed to be broken down if it jammed during an emergency evacuation, so on impact totally disintegrated, smashing into many pieces. Fragments flew everywhere, with a sizeable splinter striking SFO Atchison's shoulder, only narrowly missing his head. Had the missile struck a little higher he could have been knocked unconscious and seriously injured. As the captain lifted from his seat, his legs struck the control column, pushing it forward, and the autopilot disengaged. The 1-11's nose pitched 6° down and the wings banked 25° to the right, with the aircraft swinging 30° to the right in a diving turn. Climb power was still applied, so the speed began to build up rapidly.

Captain Lancaster's head and shoulders exited vertically through the open window but, as his chest caught the slipstream, he was blown violently backwards over the top of the fuselage. His feet pivoted forward and, as a result, his legs caught between the control column and the flight deck coaming, jamming him in position and momentarily preventing his egress through the aperture. His lower back arched over the top of the window frame with his arms outstretched with the force of the airflow. For the time being he was stuck but, with the blast of the airflow buffeting his body, he flapped wildly like a flag in a fierce wind, and he would not remain jammed for long.

At the time of the explosive decompression, Steward Nigel Ogden was working in the galley just aft of the flight deck. As the door blew off its hinges he followed not far behind, being hurled by the blast of air into the cockpit. As he picked himself up he saw to his horror the captain in the process of being sucked out of the window.

In the first few moments after the explosion the noise and chaos on the flight deck were extremely disorientating and caused great deal of confusion. SFO Atchison, however, in spite of the initial shock of the incident, managed quickly to gather his senses and swiftly to grab control of the aircraft. He was faced with a daunting task, for the 1-11 at first appeared to be badly damaged and to be spiralling out of control. Steward Ogden's reactions, fortunately, were just as fast and, also recovering quickly he threw himself at the inert pilot, grabbed him round the waist and held on as firmly as he could.

It was obvious to the rest of the crew that an explosive decompression had occurred, so the remaining flight attendants quickly instructed the passengers to fasten their seat belts, reassured them, then took up their emergency positions. There were no 'drop down' masks fitted on this aircraft and emergency oxygen was not available to the passengers, but as a precaution Steward Simon Rogers, working at the rear of the aircraft, donned the mask of a portable oxygen bottle and sat down in seat 20D. Purser John Heward was working in Club Class at the front of the aircraft and ran forward to report to the flight deck. As he entered the cockpit he was amazed at the devastation. He, too, quickly grabbed the stricken pilot's legs, hooked his arm through the seat belt of the spare jump seat behind the captain's location, and assisted Nigel Ogden in restraining the commander.

SFO Atchison, meanwhile, grappled with the stricken machine. During bi-annual simulator checks flight crew face a complete range of emergency drills, including decompression and pilot incapacitation, but a double emergency involving both was not part of the syllabus. With no other flight crew member available to help with procedures and to read checklists, Alastair Atchison was completely on his own and he would have to work from memory. Fortunately he was a cool and competent pilot. The emergency descent drill, following a decompression, required the donning of oxygen masks, closing of the throttles, pulling of the speed brakes and rapidly descending the aircraft in a diving turn down to 10,000 feet. SFO Atchison decided not to don his oxygen mask, in order to liaise with the rescuers on the flight deck. Strictly speaking, flight crew should wear masks above a cabin altitude of 10,000 feet but, with the aircraft at just over 17,000 feet and already descending, and the cabin altitude, still receiving pressurised airflow from the compressors, at just over 13,000 feet, there was little problem, and judgment would not be impaired by lack of oxygen. The second item of the procedure, however, closing of the throttles, proved to be a more arduous task, and the speed continued to build up rapidly. A large piece of the broken cockpit door lay over the central pedestal, covering the throttles and, on top of the door section, lay Steward Ogden as he continued to hold on to the captain. SFO Atchison was at first unable to reach the throttles but, with difficulty, he eventually managed to prise the door section off the levers and to close them with one hand while still holding on to the control column with the other. He was unable to get to the speed brake lever because of the debris, however, so, with the aircraft already descending rapidly, he simply held the 1-11 nose down for maximum speed. The airspeed indicator showed 340 knots, 10 knots above the maximum permitted marked by a red-and-white striped pointer on the instrument, but rather than slow the machine he maintained the achieved speed to get down as swiftly as possible. He was soon pleased to note that the 1-11 was handling satisfactorily and that the damage appeared to be limited to the missing windscreen.

'Mayday, mayday, mayday,' yelled SFO Atchison as he transmitted the distress call over the radio, 'London, this is Speedbird 5390.'

'Speedbird 5390, roger, mayday acknowledged.'

The noise on the flight deck was still deafening with the rush of air as the 1-11 dived earthwards and the reply went unheard. Also, to add to the din, outside the window Captain Lancaster still flailed in the violent wind. Cruelly pinned back against the fuselage by the ram effect of the air, his hands and head flapped and banged loudly against the cockpit's side. SFO Atchison thought his captain had been killed instantly but amazingly he was still alive. The buffeting of the icy slipstream tore all the clothes from Captain Lancaster's body, froze his limbs and crushed the air from his lungs. He could hardly breathe and he tried desperately to turn his mouth out of the airflow to gasp some air. He even tried to call out to the crew but his cries went unheard. The outside air temperature was –17°C but, at 340 knots, was much lower with the chill factor of the freezing airflow, and the severe conditions soon took their toll. After a few moments he could feel his mind going and he lapsed into unconsciousness.

SFO Atchison continued to battle with the radio, transmitting distress calls as the aircraft dived in descent, but his attempts were to no avail. The deafening noise in the cockpit drowned the controller's replies and it appeared communications were lost. Fortunately, Alastair Atchison, now recovered from the initial shock, was fully in command and thinking clearly. He realised he was on the edge of the Heathrow Airport area, plummeting downwards without air traffic control contact into one of the busiest zones in Europe, so to help avoid other traffic he turned the 1-11 back onto the original heading of 195°M. He reasoned, sensibly, that if he was clear of other aircraft in his ascent on this heading it could be safe in the descent. Once again he issued a distress call.

'Speedbird 5390, mayday, mayday, mayday, emergency depressurisation, on a radar heading of 195, descending to flight level 100.'

The sound of the rushing air and the flailing of the captain on the fuselage's side continued to impair communications and the controller's reply still went unheard. As a further precaution, the first officer decided to stop the descent at flight level (FL) 110 (11,000 feet). The level would keep the 1-11 clear of the slower traffic flying up to 10,000 feet in the congested Heathrow area.

At its maximum rate of descent, the 1-11 dropped at 4,600 feet per minute so, in under two minutes, FL110 was approached. The first officer pulled back on the controls to level the aircraft and slowly the speed began to reduce. Steward Simon Rogers, who had been sitting at the rear of the cabin, soon realised the problem was on the flight deck, and he rushed forward with oxygen bottles in case they were needed by the rescuers. Stewardess Sue Prince, now alone in the cabin, did a splendid job in looking after the passengers. With the flight deck door missing, many people seated near the front could see the drama unfold in the cockpit and they could see a body stuck halfway out of the window. The drop of the emergency descent also felt like the aircraft diving out of control and a number of passengers were in distress. Sue Prince calmly patrolled the cabin, reassuring those in her charge and placating those most in need.

On the flight deck Simon Rogers assisted with the restraining of the commander and, as he did so, Purser Heward quickly collected the cockpit door debris, pulling the large section from the centre pedestal and stowing the pieces in the forward toilet. The first officer now had access to the speed brake and he pulled the lever to slow the aircraft more quickly. He then tried to select the autopilot to relieve him from hand flying, but it stubbornly refused to engage.

As the aircraft slowed, Captain Lancaster's torso slipped to the side of the fuselage and his head and shoulders could be seen through the direct vision window. Steward Ogden, the first of the rescuers on the scene, still held on firmly with his arms wrapped round the commander's waist, but he was beginning to feel the strain. He had been exposed to the full blast of the airflow during the emergency descent and he was suffering from frostbite. One of his arms was being crushed between the captain's body and the window frame and, with the buffeting of the wind, was knocked against the metal causing cuts and bruises. He was in a great deal of pain and would be unable to hold on much longer. Captain Lancaster's legs were still caught up in the controls, with his right leg jammed between the control column and the cockpit coaming and his left leg bent back and wedged against the seat cushion. The other two rescuers managed to grasp the commander's legs and Nigel Ogden, his strength almost gone, was able to release his grip from round the inert pilot's waist. Hurt and in pain, he retired to the cabin to recover from his wounds.

The first officer was still unable to select the autopilot and, on checking the equipment on the central pedestal, discovered that in the mêlée after the decompression the autopilot master switches had been knocked off. He then selected the switches to on, engaged the autopilot, and once again attempted radio contact.

'Speedbird 5390 is maintaining 110.'

Still nothing was heard. SFO Atchison switched to very high frequency (VHF) box 2, which was tuned to the company frequency at Birmingham, and transmitted once more. A somewhat confused voice in the BA office at the airport responded to the call but, rather than reply, the first officer switched back to VHF1. It was then that he noticed that the VHF1 switch on the central pedestal had also been knocked off and reselecting it to on he called again that he was level. This time his transmission was received but with the noise in the cockpit he was still unable to hear the response.

As the aircraft slowed still further the two remaining rescuers attempted to pull the captain back into the flight deck but they were unable to succeed. The slipstream continued to pin his body against the fuselage side and he was impossible to move. With the commander stuck out of the window, Alastair Atchison had little choice but to land as quickly as possible. He decided that, if communications could not be re-established soon, he would simply divert straight to Gatwick, an airport with which he was familiar. He now prepared the aircraft for an immediate landing. It was required to set the airspeed bugs for flap extension and landing speed but, since the data had blown out of the window, SFO Atchison had to work from memory. He was aware, however, that at the same aircraft weight the airborne safety speed with take-off flap set, V2, was close to the minimum permitted reference speed over the threshold for landing with landing flap set, Vref. Since the weight was much the same as at take-off, he simply based Vref and the required flap extension speeds on the V2 speed which was still bugged on the airspeed indicator. As the 1-11's speed dropped below 220 knots the speed brake was stowed, the first stage of flap was selected and, as the speed decayed further, more flap sections were extended. With the speed decaying the noise level reduced and those on the flight deck were at last able to confer.

'We think he's dead . . . pretty sure he's dead', shouted one of the rescuers to the co-pilot.

The London controller, meanwhile, was still attempting to establish contact with Speedbird 5390, making several calls without success. Unaware of the precise nature of the 1-11's emergency, he continued to operate the sector as normal, communicating in turn with several flights. Some of the crews tried to relay on the radio with the stricken aircraft but their efforts were also thwarted.

By now Simon Rogers was holding both Captain Lancaster's legs on his own so, with the aid of Purser John Heward, he was strapped into the jump seat for security. As he manoeuvred into position, however, Steward Rogers suddenly felt the commander slip. For a moment the rescuer's heart missed a beat as he thought the inert pilot had gone, but fortunately he only slipped about six inches more out of the window and he was able to restrain him from falling further. The entire top section of Captain

Lancaster's body now flapped out of the window with only his legs inside, but Simon Rogers, from his position strapped into the seat, could now hold on to him tightly. With the captain secure there was little more for the purser to do so he left the flight deck to check the injured steward and to help Stewardess Prince look after the passengers.

As the 1-11 slowed below 170 knots the landing gear was lowered and, when locked down, the remaining two stages of flap were selected. At 160 knots power was applied and the speed stabilised. The aircraft was now ready for an immediate landing and Alastair Atchison completed the checks from memory and made his final preparations. With all checks completed and the landing gear and flaps set, he was now free to concentrate on the flying without interruption. It was very sensible to get these items out of the way for he would have enough to cope with on his own. Another attempt was made to establish communications and, after nearly seven minutes of radio silence, he was able to hear the controller's voice and establish communications.

'We have emergency depressurisation, requesting radar assistance, please, for the nearest airfield.'

'Speedbird 5390, roger,' replied the London controller, 'can you accept landing at Southampton?'

'Speedbird 5390, I am familiar with Gatwick, would appreciate Gatwick.'

The controller then cleared the 1-11 direct to Mayfield, a radio beacon lying 13 nm south-east of London's Gatwick Airport, but on receiving the instruction SFO Atchison had second thoughts. It was imperative to land as quickly as possible and Southampton was obviously nearer. It would also be quieter with less traffic so, although the co-pilot had never been there before and was unfamiliar with the airport, he decided to accept the controller's advice.

'Speedbird 5390. I am maintaining level 110 and I am at 150 knots, requesting radar assistance into Southampton.'

The 1-11 was cleared to descent to FL70 and to call London's Southampton sector controller on frequency 134.45. The descent was commenced but, in the confusion on the flight deck, the frequency change was not made, and the Bristol sector controller could still be heard issuing instructions to a number of aircraft. SFO Atchison called back saying he was descending out of FL80 for FL70, and it was then realised that he was still on the same frequency.

'Speedbird 5390, roger, remain on this frequency then, sir, and I will give you radar vectors into Southampton.'

The first officer now spoke to the passengers and assured them the aircraft was under control. He informed them they would be landing shortly and asked them to follow the instructions of the flight attendants. In the cabin, preparations were made for landing and the passengers refreshed on the emergency evacuation procedures. Outside the window, Captain Lancaster, still flailing in the wind, was taking a lot of punishment, but there was nothing more that could be done. His legs were still impeding the control wheel but, with its movement as SFO Atchison flew the aircraft, the limbs were beginning to disentangle. Steward Rogers managed to free the commander and to hold him round the ankles, thus releasing the controls. Simon Rogers, now the only link between rescue and certain death, held on grimly.

BA5390 was cleared further descent to 4,000 feet but once again the instruction was missed. The environment on the flight deck was still noisy and chaotic and Alastair Atchison was kept very busy conducting the flight, operating the radio, preparing for the approach and landing and liaising with the rescuer and the purser. London's Bristol controller checked once again if his message could be heard and the 1-11 replied.

'Roger, reading you, er, strength five. I'm afraid we have debris in the flight deck and, er, can you confirm the frequency you changed me to.'

'Okay, sir, if you remain on this frequency and continue down to 4,000 feet, please.'

The controller omitted to include the Southampton local area pressure setting (QNH) with the descent clearance, so when the first officer requested the information, he instructed the 1-11 to change frequency for the details.

'And, Speedbird 5390, if you check that now on, er, frequency 131.0, Southampton Approach.'

SFO Atchison changed frequency and called the Approach controller who immediately took over positive control of the flight.

'Speedbird 5390, good morning,' replied Approach, 'identified on hand-over London radar, six miles to the west of Southampton Airfield. What is your passing level?'

'Roger, sir, presently leaving flight level 64 (6,400 feet), could you confirm, er, your QNH, please.'

'Roger, my QNH 1019 millibars (mb), the runway in use is runway 02, the wind is 350/12.'

'Roger, sir, I am not familiar with, er, Southampton. I request you shepherd me onto the runway please.'

'Roger, that is copied. Roll out, then, onto a heading of 180.'

The 1-11 was being radar vectored from the west of the airport in an anti-clockwise direction on to the north-easterly runway, and was further instructed to turn left from its southerly heading onto a heading of 110. The Approach controller then requested the number of passengers on board, a figure required by the emergency services, and asked if the pressurisation failure was their only problem.

'Er, negative, sir, the, er, the captain is half sucked out of the aeroplane. I understand . . . I believe he is dead.'

'Roger, that is copied.'

'Er, flight attendant's holding on to him but, er, requesting emergency facilities for the captain. I . . . I think he's dead.'

'Roger, that is copied,' confirmed the controller.

'Continue your descent, then, to 2,000 feet, QNH 1019. Make it a nice gentle turn at the moment . . . you're seven miles south-west of the airfield.'

Passing 5,500 feet the 1-11 was radar vectored back onto a southerly heading to permit a more gentle turn for the approach. The first officer then requested assurance that Southampton was suitable for a 1-11, saying that he'd prefer a runway length of 2,500 metres. The runway length was relayed back as being only 1,800 metres, but that was confirmed as acceptable. Descending through 3,900 feet, BA5390 was further cleared to 1,700 feet, and it was agreed that a 20-mile final would give the co-pilot sufficient time to settle for the approach. At about 3,000 feet, SFO Atchison started the auxiliary power unit (APU) in preparation for shutting the engines down once they had stopped on the runway, and then called the purser to the flight deck to tell him to expect a normal landing. Steward Rogers, still holding on to the commander as tightly as possible, suddenly felt the captain try to kick his legs. Could Captain Lancaster, after so much punishment, still be alive after all?

'5390,' called Approach, 'commence a gentle turn now, onto a heading of 360. I'll give you twenty track miles to run for touchdown.'

The first officer then requested the instrument landing system (ILS) frequency, but was informed that only a VOR (VHF omnidirectional radio range) beacon was available. An ILS would have provided both centre line and glide path guidance, but using the VOR only centre line guidance would be available. To add to the co-pilot's problems he now had to fly a more difficult approach without glide path information. Fortunately Alastair Atchison was handling the drama well and his skills were equal to the task.

Having established the availability of only a VOR approach, the controller assured the 1-11 that radar vectors would be given to final approach.

'5390, thank you very much,' replied SFO Atchison, 'we are three greens, er, and flaps 45, so I'm set for an approach, but make it, please, very gentle.'

'Yes, I will do, indeed, you are number one in traffic.'

Settled on a heading of north and now level at 1,700 feet on the airport pressure setting (QFE) of 1017 mb, the aircraft was instructed to turn right onto a heading of 025° for final approach.

'Speedbird 5390 is nine miles from touchdown. You are cleared to land. The wind indicates 020/14. Descend to height 1500 on QFE 1017.'

'Roger, sir, descending to 1,500 feet. Talk me down all the way . . . I need all the help I can get.'

The Approach controller was doing an excellent job as well and was also equal to the task. He assured the co-pilot of his assistance and confirmed that the 1-11 would stop on the runway and that emergency services were standing by. On the flight deck the atmosphere was tense and SFO Atchison required all his concentration. Steward Simon Rogers still held on tightly to the captain's ankles.

Outside the visibility was hazy and the first officer could still not see the runway, so the controller issued a series of instructions for guidance on final approach.

'Your range is now seven miles from touchdown. You're on the extended centre line.'

'Your range is now six and a half miles. You are cleared to land. You are on final approach track.'

'5390, turn left 5 degrees. You are five miles from touchdown.'

'Your range is four miles, your height should be 1,250 feet and the wind is 020/10.'

'Three and a half miles from touchdown, turn right 3 degrees. On final approach track, heading is good.'

'You are 3 miles from touchdown, the height should be 950 feet on 3 degree glide path. You are lined up. You are cleared to land.'

'5390, thank you,' called the co-pilot, 'I have the runway in sight.'

'You are cleared to land,' confirmed the controller once more, 'do you wish me to continue with further information?'

'Negative.'

The first officer continued the approach to finals visually and, judging it well, crossed the threshold at the appropriate height and speed and executed a smooth landing.

'Speedbird 5390, fantastic approach,' radioed the controller, 'you may shut down on the runway and leave the frequency.'

The 1-11 was quickly slowed on the runway and at 08:55, twenty-three minutes after the explosive decompression, was brought to a stop. Electrical power was swiftly transferred to the APU, the engines were shut down and, as the fire and rescue services attended to the captain, the passengers were disembarked by airstairs from the front and rear doors.

A fire engine had to be backed up to the nose of the 1-11 to use as a platform before the commander could be lifted and manoeuvred through the window back into the cockpit. As he was being helped, Captain Lancaster, to everyone's surprise, suddenly regained some consciousness and asked how high they were.

'About four feet six inches, mate,' replied one of the firemen, 'but don't worry about it.'

By some miracle Tim Lancaster had survived. He was still semi-conscious, seeing red-and-white blurs of fire engines and ambulances and mumbling to himself, but he was very much alive. He was rushed to hospital where he was found to be suffering from fractures to his right arm and wrist, a broken thumb, bruising, frostbite and shock. After many months of pain and discomfort he went on to make a brave recovery and is now back flying with BA on the 757/767 fleet. But for the ability, resourcefulness and courage of those involved it could have been another story. Captain Lancaster praised his co-pilot for a 'magnificent performance in bringing the aircraft down safely' and his rescuers for their 'courage and bravery' in saving his life. The official report of the incident stated that 'the fact that all those on board the aircraft survived is a tribute to the crew's quick thinking and perseverance in the face of a shocking experience.'

The airmanship displayed by Alastair Atchison was of the highest order and the awards he received were a true testament to his skills: the Queen's Commendation for Valuable Services in the Air, the Guild of Air Pilots and Navigators Master's Medal, the Hugh Gordon Burge Trophy for Safety, the British Airline Pilots' Association Gold Medal, the International Federation of Airline Pilots Association Polaris Medal, the Federation Aéronautique Internationale Diploma for Outstanding Airmanship, and the British Airways Award for Excellence.

The bravery of the cabin crew was no less rewarded, and all four, John Heward, Nigel Ogden, Sue Prince and Simon Rogers, received the Queen's Commendation for Valuable Services in the Air and the British Airways Award for Excellence.

The entire crew, including Tim Lancaster, also received the 'Man of the Year' Award. Stewardess Sue Prince, the first woman to receive the accolade, was made an honorary man for the occasion.

Chapter 3

Pacific Search

The vast Pacific Ocean is bounded by five continents and covers an area of seventy million square miles. Its deep waters stretch from the shores of Asia and Australia to the coastlines of North and South America, and to Antarctica in the south. Along the line of the Equator from Indonesia to Equador the distance across the waves is over eight thousand nautical miles (nm).

The name, Pacific, belies the temperamental nature of the ocean, and on occasions storms of great ferocity lash the region. Even large modern airliners have to treat the area with caution, for all but very-long-range aircraft must island-hop to overcome the immense distances. Fortunately, the surface of the Pacific is dotted with thousands of islands, although they are mostly very small and widely scattered. The only regular flights to close the enormous Pacific gap are the 747SP (Special Performance) non-stop services of United Airlines and Qantas from Sydney to Los Angeles, a journey of 7,475 miles which takes thirteen and a half hours to complete.

Aboard modern passenger jets, electronic navigation equipment, adequate fuel reserves and skilled crews have transformed flights between far-off islands to safe, routine and everyday events. For the light aircraft pilot, however, the Pacific still presents an awesome barrier. The journeys from island to distant island are long and tedious, with flight at slow speed. Navigation over the featureless stretches of sea is difficult, and any error or malfunction in locating the remote specks of land can prove disastrous. For one pilot venturing alone in a light aircraft across the vast pacific in 1978, a simple instrument failure led to the aviator becoming hopelessly lost.

In 1978 Jay E. Prochnow was thirty-six years old and a very experienced pilot who had flown in the US Navy. He had also completed a tour of duty in Vietnam. At the time he worked as a delivery pilot for Trans Air of Oakland Airport, California. One assignment he had been given was to fly a Cessna 188 – a light, single-engine, single-seat aircraft used for crop spraying – from California across the Pacific to a customer in Australia. Such long-distance flights, although hazardous for a small machine, are cheaper than inter-continental shipping costs. Extra fuel tanks were fitted to the aircraft for the ordeal. The only sources of navi-

gation, however, apart from chart and compass, were the rather anti-
quated radio beacons known as non-directional beacons (NDB) scattered
amongst the islands of the Pacific.

Although these beacons are relatively inaccurate for the distant plotting of position on a chart, some of the more powerful transmitters have
signal ranges in excess of 300 nm over the sea by day and 700 nm by night.
These radio beacons transmit a signal, not unlike a broadcast station, and
Prochnow's aircraft was fitted with an automatic direction finder (ADF)
which could be tuned to a particular NDB and the transmitter identified
by its morse code emission. The ADF could then detect the incoming
signal and point a needle in the direction of the beacon. The needle direction and compass reading could be utilised to establish the bearing of the
aircraft from a distant beacon for plotting a position line on a chart, and
three or more bearings could be taken from different beacons to fix the
position roughly. At long ranges the needles of the ADF were known to
fluctuate wildly, and precise track-keeping was very difficult using such a
system. It was sufficiently accurate, however, to guide the aircraft to
within range of the destination NDB and the pilot could then home in on
the signal by simply following the direction of the needle.

A few days before Christmas, 1978, Prochnow arrived in Pago Pago in
the American Samoa Islands, accompanied by a colleague in another
Cessna 188 who was flying the same journey. By then they had successfully completed well over half the trip and if all went according to plan
they would arrive in Australia in time to return home for Christmas. Two
days later the two Cessnas took off together at 03:30 local time with
Prochnow in the lead. Unfortunately the second aircraft's fuel pump shaft
sheared at lift-off and Prochnow watched with horror as his colleague was
forced to ditch in the sea. Fortunately the downed pilot escaped unhurt.
Prochnow returned to land and after a day's rest left on his own with full
tanks in the middle of the following night for a now lonely and gruelling
trip to tiny Norfolk Island, 1,475 nm away. Cruising at speeds of around
110 knots, the journey would take a minimum of fourteen hours to
complete in still wind conditions, but would be nearer fifteen hours flying
against the forecast light westerly winds. The flight time would allow a
daylight landing in Norfolk in mid-afternoon at about 16:00 local time,
giving sufficient leeway for adverse winds. The full tank fuel load gave a
total endurance of twenty-two hours at normal cruise speeds, which
seemed an adequate reserve for any contingency. If necessary fuel could
be conserved by decreasing power a little and by slightly leaning the
mixture, i.e. reducing the fuel-to-air ratio, thereby increasing the
endurance.

Norfolk Island is situated at 29°S 168°E, about 800 nm off the east
coast of Australia. The island lies about 600 nm NNW of Auckland, New
Zealand, and 430 nm south of Noumea, New Caledonia. It would not be

inaccurate to describe Norfolk Island as a tiny speck of rock lying in the middle of nowhere in a corner of the vast Pacific. There is no land around for over 400 nm in any direction. Although Norfolk Island is Australian territory, for convenience it is enclosed within the Auckland Oceanic Flight Information Region, and all flights in and out of the island airport communicate with the Auckland Air Traffic Control Centre (ATCC).

Prochnow took off from Samoa in the darkness at about 03:00 local time on 21 December and climbed slowly under the weight of fuel to his cruising altitude of 8,000 ft (2,438 m). He turned due south-west to pick up the direct track of 220° magnetic from Pago Pago to Norfolk Island and settled down for the long flight ahead. With no autopilot to ease the monotony of hand flying, it was going to be a tiring and tedious journey. There would be no chance of a quick nap. He would have to navigate with a map on his knee and plot position while flying the aircraft, and would have to grab snacks from his pack meals whenever he could. The initial routeing lay over the Tonga, or Friendly Islands, then continued on southwest bound about 200 nm south of Fiji. For most of the first part of the route sufficient NDBs were available for reasonably accurate fixing of position and progress was good.

A few hours later the strain of flying through the blackness of the night was eased when dawn broke behind the tail of the Cessna at about 05:15 local, and bathed the ocean in light. Approximately 600 nm from Pago Pago lay the island of Ono-I-Lau, almost directly on route. Prochnow navigated toward the island using the various NDB beacons along the way to plot his track. In due course Ono-I-Lau was sighted and his position was confirmed. Beyond the tiny island stretched a vast area of ocean with the destination still over 850 nm away. Prochnow would now have to traverse several hundred miles of empty sea, without any source of navigation whatsoever except the basic magnetic compass, before picking up the signal of the powerful NDB at Norfolk Island. He could then home in on the beacon over the last two or three hundred miles and be guided to the local airport.

At almost the halfway point of the journey the little Cessna crossed the international date line at 180° E/W longitude, the line at which time is both twelve hours behind GMT and twelve hours ahead of GMT, i.e. at the same time on each side of the line, but on different days. As Prochnow traversed the date line he lost twenty-four hours and jumped one day ahead. The local date was now 22 December 1978 and the local time around 08:00. Approximately eight hours of flying remained.

A further 120 nm along track the small machine crossed the Tropic of Capricorn, the line on a chart marking 23½° South, the most southerly declination of the sun. This day the sun was at its highest point in the southern sky, for 21 December marks the Winter Solstice, the shortest day in the Northern Hemisphere's winter, and the longest day of the summer in the south. Prochnow's trip was well planned, for the maximum amount of daylight would be available for the long flight to Norfolk Island.

After several more hours of flying over the featureless sea the Cessna came within range of the destination radio beacon and the signal was identified by its 'NF' morse coding. The arrows of the ADF pointed directly ahead. Prochnow was now flying within the jurisdiction of the Auckland Oceanic ATCC and had established contact on high frequency (HF) long-range radio. His estimated time of arrival (ETA) at Norfolk Island was given as 04:30 GMT, 16:00 local time.

The little Cessna droned on at 8,000 ft (2,438 m) as Prochnow followed the direction of the ADF needle. With the approach of the estimated time of arrival he peered ahead for a glimpse of the tiny island, but the bright sun made forward vision difficult. The ETA came and went with no sign of Norfolk, but the ADF needle still pointed steadfastly ahead. The winds could very well have been stronger than forecast, which would easily affect the flight time, although arrivals were normally within fifteen minutes of ETAs. It was just as well Prochnow had loaded plenty of fuel. He called Auckland and informed them of the situation but as yet felt no cause for alarm. Thirty minutes later the needle still indicated the island lying ahead. The bright sun began to fall in the western sky, further reducing forward visibility as Prochnow's eyes searched eagerly for a sight of the

Captain Gordon Vette.

land. He tuned in two distant powerful radio beacons, one on Lord Howe Island and the other at Kaitaia in northern New Zealand, in an attempt to plot his position, but the resultant fix was right off his chart. Something was seriously wrong. On retuning the Norfolk beacon he saw to his horror that the needle now pointed in a completely different direction. In spite of the Norfolk Island DNB being correctly tuned and identified the functioning of the ADF appeared to be seriously amiss and the instrument seemed simply to point at random in any direction. Prochnow rapidly gathered his senses and assessed the situation. He had about seven hours of fuel remaining, perhaps more if he conserved his fuel. What should he do now? Which way should he turn for Norfolk Island? For some time he had been flying in the direction of the ADF needle which pointed where it chose. At this stage he could be sixty miles (ninety-six kilometres) or more adrift from his track to destination. He had no doubt that by now his situation was desperate. Prochnow radioed Auckland declaring an emergency and gave details of his predicament. He was hopelessly lost. Alone in a small aircraft in the middle of the empty Pacific, and with sunset only a few hours away, his chances of being found were very slim indeed. Immediately he began a square search of the area. If, while looking for Norfolk Island, his tanks ran dry he would be faced with ditching at night in the lonely Pacific, a prospect which was not appealing. If he was forced into such a situation his chances of survival would be nil.

At about 17:15 local Fiji time, Flight TE 103, an Air New Zealand DC-10, registration ZK-NZS (Zulu Sierra), took off from Nadi airport for Auckland, some three hours' flight time away. The captain, Gordon Vette, a senior pilot with the company, turned the jet due south for New Zealand and climbed the aircraft to its cruising altitude of 33,000 ft (10,060 m). His co-pilot on the flight was First Officer Arthur Dovey and his flight engineer, Gordon Brooks. About twenty minutes later, as the big jet settled at cruise altitude for what was a short hop for the long-range aircraft, news of Prochnow's predicament came through from Nadi on very high frequency (VHF) short-range radio. Further information could be obtained on HF from Auckland ATCC who were handling the Cessna's flight.

By the time contact was established with the Centre, the news from Auckland was bleak, and it was obvious to the crew that the situation was very serious. Prochnow was completely lost and his reserves of fuel were being steadily consumed. Even if help were available immediately it could be a tight-run race for safety. In the busy air and shipping routes of the western world, maritime services and air-sea rescue units would be quickly mobilised to help, but in the remote Pacific it is not such a simple matter. An Orion aircraft of the RNZAF had been placed on standby at Whenuapai Air Base, twenty miles (thirty-two kilometres) north of Auckland, but it would be some hours before it could reach the search area once it became airborne. The Air New Zealand DC-10 would pass about 400 nm due east of Norfolk Island on its route from Fiji to Auckland, but if it proceeded directly to the search area it could be there in about one and a half hours. Since there was very little other aircraft activity in the region, would the crew of the New Zealand jet be able to help?

Captain Vette and his colleagues did not need to be asked twice, and soon the big jet was speeding towards Norfolk on its rescue attempt. By good fortune the DC-10 carried a lot of extra fuel and could remain airborne for some considerable time. In New Zealand aviation fuel is expensive and the jet was 'tankering' fuel to save costs. The flight was also scheduled to proceed to Wellington and the extra fuel load would save refuelling in the transit. But there was more in Prochnow's favour than he might have hoped, for Vette, although a senior captain, was an enthusiastic navigator and still kept his flight navigator's licence current. The DC-10, of course, had sophisticated electronic navigation units on board consisting of three completely separate area inertial navigational systems (AINS), and did not require a navigator, but the DC-8 aircraft in Air New Zealand's fleet at the time were not so electronically equipped. All co-pilots on such fleets were also trained as navigators and performed navigation duties as and when required. Vette used to navigate occasionally on the DC-8 to keep his licence up to date.

The AINS of the DC-10 could navigate with pin-point accuracy and displayed a continuous read-out of the aircraft position at all times. There was no equipment aboard, however, for homing in on the Cessna, and trying to find the light aircraft would be like looking for a needle in a haystack. The radar on the DC-10 flight deck was used only in scanning for weather and could pick out large storm clouds, but would never receive an echo from such a small machine. In this situation the specialist knowledge of the navigator could be put to good use. While the crew continued to co-ordinate the rescue plan with Auckland, Vette turned his attention to the passengers. All on board were expecting a short flight, and since the search could be a lengthy procedure the arrival in Auckland could be quite late. Fortunately there were only eighty-eight passengers. With a life at stake there would be none who could object to the crew's action.

'Ladies and gentlemen,' spoke Captain Vette from the flight deck, 'we've just received news from Auckland of a light aircraft lost in the region of Norfolk Island and since we're the only suitable aircraft in the vicinity we've altered course for the search area to offer assistance. The pilot is in serious danger of ditching and if he lands on the water his chances of survival will be remote. The search could be lengthy, but fortunately we have plenty of fuel. It could mean a rather late arrival in Auckland, however, but with a life at stake I'm sure you understand that all help should be given. If anyone objects, of course,' added Captain Vette jokingly 'we could always leave him to die!'

Meanwhile the DC-10 called Prochnow on a range of HF frequencies and finally managed to establish contact. The Cessna pilot was told of their intentions and that they would do all they could to assist. It was a relieved man who received news that help was on the way for Prochnow was by now three hours overdue on his original arrival estimate. The DC-10 crew were informed that his ADF appeared to be malfunctioning and that only four hours of the Cessna's fuel remained. In spite of his predicament Prochnow spoke on the radio with calmness, and his composed demeanour impressed the listeners. He was a very cool customer indeed.

Amongst the passengers on board the DC-10 was an Air New Zealand first officer, Malcolm Forsyth, like Vette a qualified and still licensed navigator. After the public address (PA) announcement to the passengers he came forward to the flight deck to offer his help. It was amazing that two current navigators should be on board, and the good news was transmitted to Prochnow. The two navigators could co-ordinate the search while First Officer (F/O) Dovey and Flight Engineer (F/E) Brooks could fly the DC-10 and monitor the aircraft systems. All crew members would be required at times to operate the radios for a lot of co-ordination work would be required.

As the DC-10 flew southwards Vette questioned Prochnow on all the details of his flight – altitude, airspeed, estimated fuel consumption, estimated fuel remaining, estimated position, if he had any idea – anything they could grasp which might improve the chances of finding him. All communications were being conducted on HF long-range radio, which is subject to static and background noise, and communications proved difficult. The level of noise in the Cessna cockpit was high and did not help the situation. As the flight continued, Vette still had the needs of his passengers in mind and they were kept fully informed and up to date on the progress of the search.

'When we reach the search area,' added Captain Vette 'you will all be able to help. The more pairs of eyes searching the skies the better the chances of finding the Cessna.'

Vette encouraged his passengers to feel part of the search team and little groups were invited in turn to the flight deck to witness the proceedings.

The next problem facing the DC-10 crew was to try and establish the position of the Cessna. A specialised navigation plotting chart was needed but none was carried on the DC-10. Once again luck was on Prochnow's side, for Vette searched his briefcase and found a spare navigation chart of the region. The Cessna pilot's fortunes were beginning to change. Captain Vette instructed Prochnow to take a series of bearings from what the experienced navigator know to be powerful NDBs on the north island of New Zealand: Kaitaia on the most northern tip, Tauranga on the coast just north of Rotoru, and Gisborne on the east coast. Although Prochnow's ADF was suspected of being in error, a positive test of the integrity of the equipment was essential. Vette carefully plotted each reading on the chart but the result was nonsense, for it showed the Cessna to be somewhere south of Auckland! There was no doubt now that Prochnow's ADF was malfunctioning and would be useless in the search.

The Cessna's equipment was able to tune and identify radio beacons but unknown to anyone at the time the problem was caused by a most elementary fault. The needle of the ADF had simply become loose on the spindle and pointed totally at random in whichever direction it settled. Also unknown was the fact that the defect had taken the Cessna 200 miles (320 kilometres) to the south-east of Norfolk Island and that only by some miracle could Prochnow now be saved. That, however, did not take into account the determination of Captain Vette and his crew.

The next attempt at establishing position was made using long-range HF direction-finding (DF) equipment at Brisbane. The DF system is normally employed using VHF radio. An operator at an airport receives from a flight an incoming radio message from which equipment can detect the direction of the aircraft. This magnetic bearing information is then passed to the pilot for plotting on his chart. It is a useful and accurate method of establishing position. Using HF, DF is a totally different

An Air New Zealand DC-10. *(John Stroud)*

proposition, with radio messages being transmitted for perhaps thousands of miles. Several attempts at obtaining a bearing from the HF station at Brisbane while receiving Prochnow's radio messages gave a rough indication of direction, but a more accurate method of establishing position was required if he was to be found.

The DC-10 and the Cessna were still communicating on the rather difficult HF, but it was expected that shortly the two aircraft would be within VHF range. Prochnow was asked to call repeatedly on the VHF emergency frequency of 121.5 MHz to establish the exact point of contact. Since the maximum range of VHF transmission is of the order of 200 nm, at that juncture the distance of the Cessna from the DC-10 would be known. It would still place Prochnow at the centre of a circle of 200 nm radius, covering an area in excess of 166,500 square miles (323,750 square kilometres), but at least it would be a start. Since the position of the Cessna was unknown, it was considered essential to attempt to establish the relative positions of both aircraft. It could be that Flight TE 103, instead of speeding towards a rescue at Norfolk Island, was flying away from the light aircraft. Once again Vette's expertise was put to the test. The sun was setting low in the sky with the approach of dusk and its direction might just prove useful as a last resort.

'I have an idea,' called Vette over the radio above the noise of the HF static, 'turn due west and face into the sun. When you are pointing directly at it read me back your compass heading.'

After a few moments came Prochnow's response: 'Heading two seven four degrees', he radioed.

The large DC-10 also aimed straight for the sun and settled on a heading of 270°. That placed the Cessna, as expected, to the left, i.e. to the south of DC-10 'Zulu Sierra.' But was it east or west of the jet's position? A sextant on each aircraft could have measured the precise angular altitude of the sun above the horizon and the difference in altitudes could have been used to clarify the situation, but without such apparatus it was difficult to judge. Vette, however, recalled that a clenched fist held vertical at arm's length represented about ten degrees above the horizon and a finger a little more than a degree and three-quarters.

'Hold your arm at full length,' Vette instructed Prochnow, 'and measure the distance between the horizon and the centre of the sun using your fist and fingers.'

'That's impossible,' came the quick reply from the Cessna, 'the cockpit is small and with the windshield so close I can't even stretch my arm halfway out!'

It was a short moment of amusement that swiftly passed. As a compromise each agreed to hold his hand about one foot (thirty centimetres) before his eyes and to count the number of fingers.

'I make it almost four fingers', radioed back Prochnow.

Vette's reading was almost two and a half fingers thick, which placed the Cessna nearer the sun, i.e. to the west of the DC-10. The difference in measurement of about two and a half fingers placed the altitude readings at just under three degrees apart. When a celestial body is observed from two separate points it can be shown mathematically that the difference in measured angular altitudes expressed in units of minutes of arc is equal to the distance in nautical miles separating the observers' positions. Three degrees is equivalent to 180 minutes of arc, which, with the reading a little more than that, placed Prochnow's position approximately 200 nm away. It was now established that the Cessna was to the south and west of the DC-10. If the calculations were correct the two aircraft should be in VHF contact very shortly. A few minutes later F/O Dovey picked up Prochnow's voice on the VHF emergency frequency of 121.5 MHz.

'I've got him,' shouted Dovey, 'he's coming over clearly now on VHF.'

The efforts of the crew were beginning to bear fruit and the co-ordinates of the first point of radio contact were accurately plotted on the navigation chart. At this busy moment in the flight the chief purser, Paul James, was asked to liaise with the passengers and to pass on the good news over the PA. His announcement was greeted with cheers. Vette now instructed Prochnow to turn eastwards towards them with his tail to the sun while Zulu Sierra flew towards the sun. They should now be flying directly at each other and closing the gap more quickly. With the DC-10 flying at a ground speed of 560 knots and the Cessna at 110 knots they should cover the 200 nm separation with ease.

'We'll be directly over Prochnow in eighteen minutes', declared F/O Forsyth, the DC-8 co-pilot and navigator. Vette concurred. The next problem for the two aircraft was to spot each other, not an easy prospect with the sun low in the sky. Captain Vette had already encouraged his eighty-eight passengers to keep guard at the windows and he now instructed them in proper look-out techniques.

'Move your eyes only one or two inches along the horizon, then stop. Move another inch or so and then stop again, always moving in short bursts. Move down a space then repeat the sequence back across your vision, and so on.'

The Cessna was much lower than the DC-10, and using this method a full square search of the sky in view to each passenger could be made. It might just enable one of them to spot the tiny aircraft.

The DC-10 crew now turned their attention to making their own presence more conspicuous to Prochnow. Dovey turned 'Zulu Sierra' to see if they were forming a vapour trail for the Cessna to spot but nothing was in sight. The Met office in Auckland was contacted and confirmed that in the present conditions the formation of a contrail was unlikely, but it was worth a try. The first officer climbed, then descended the aircraft to different altitudes in an attempt to induce a trail, but to no avail. A few trips previously F/E Brooks recalled witnessing an aircraft dump fuel and how impressed he had been with the sight. Perhaps, suggested Brooks, if the conditions were right, dumping fuel would leave a bright trail which would help Prochnow spot them more easily. At each wingtip, nozzles pointed rearwards and valves could be opened to allow pumping of fuel overboard at the rate of 2.5 tonnes per minute. Prochnow, in fact, had used the same trick in the Navy as a tanker pilot to help receiver aircraft spot the tanker.

At the rendezvous time Brooks threw open the fuel jettison switches and fuel poured from the nozzles. Two minutes later he shut the valves; five tonnes of fuel had gone!

'Can you see us?' called Vette, 'you should be able to see our fuel dump trail.'

Prochnow replied, disappointedly, that nothing was in sight. Perhaps they could try again. A second time Brooks opened the valves and on this occasion dumped fuel for three minutes. Over seven tonnes of fuel spilled into the sky, leaving a distinct trail about thirty miles (forty-eight kilometres) long. That made a total of more than twelve tonnes of fuel dumped. With bated breath the DC-10 crew waited for Prochnow's call of a sighting but nothing was heard. The minutes passed and then came Prochnow's voice over the radio.

'It's hopeless,' he said, 'I can't see anything.'

The light was now fading and it was a bitter blow to the searchers that their efforts had been thwarted.

With darkness approaching Prochnow resigned himself to ditching in the Pacific for it now seemed that any attempt at finding him was going to fail. Carefully he prepared the inflatable life raft and emergency food and water supplies for survival at sea, and made sure all sharp objects were removed from his pockets. He had been in the air for nineteen hours and the strain was beginning to tell. He still had about three hours of fuel remaining, perhaps more, since he had been flying at economy cruise for some time with reduced power and leaned mixture, but at almost five hours overdue on his original arrival estimate he could by now be hundreds of miles from Norfolk. Ditching seemed almost a welcome relief to Prochnow as he was so tired and he had to fight to keep himself going.

Vette and his crew were not prepared to give up, and a further attempt at establishing position was begun using a technique known as 'aural boxing'. Prochnow would be required to make repeated radio calls, as he did when initially establishing VHF contact, and the points of lost communication and renewed radio contact would be recorded and plotted on the chart. The idea was to fly in a straight line, in this case at first roughly westbound, through the area of the circle of VHF radio

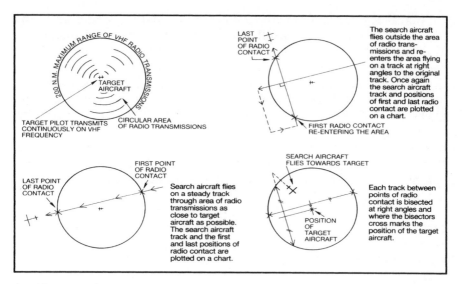

Aural boxing technique.

transmissions from the Cessna formed by the sweep of the 200 nm radius of the maximum range of Prochnow's radio. The longest line through the circle would, of course, be the 400 nm diameter, taking over forty minutes to traverse, and since the DC-10 was likely to be fairly close to the Cessna, i.e. the centre of the circle, it could be a time-consuming process. The track towards the light aircraft had already been plotted, however, and the point of original radio contact had been marked on the chart, so time could be saved by re-establishing the original track and continuing the traverse. This meant, of course, flying away from Prochnow's estimated position, not an easy decision to make, but since until now all attempts at visual contact had failed it might just be worth a try.

When the traverse was completed and the point of lost VHF contact noted, the plan was then to sweep south-east bound outside the circle of VHF range and then to turn roughly due north to repenetrate the circumference flying at right angles to the original track. Once again the points of radio contact and lost communications would be recorded. The position of the Cessna would then be established by bisecting the lines of traverse and by noting the point at which the bisectors crossed. The Cessna's position was believed, with some justification, to be somewhere south-east of Norfolk Island, so while the 'aural box' method was being applied Prochnow was encouraged to fly in a north-westerly direction to help close the gap. Pin-pointing the Cessna's position would be more difficult with a moving target, but with time and fuel running out it was better than having Prochnow fly round in circles.

Something more immediate was required to try and establish Prochnow's range from Norfolk and once again the navigators conferred as to the best solution. The sun was almost setting over the Cessna and so Prochnow was asked to record the precise GMT time of sunset, i.e. the time at which the upper limb of the sun's circumference dipped below the horizon. Norfolk Island personnel were also asked to record their own GMT time of sunset, which was about 19:00 local, and the two times were compared. A small correction had to be applied to Prochnow's observation, for at 8,000 ft (2,438 m) sunset would be seen a little later than at the same position at sea level. When the calculations were complete the separation in sunset times was found to be twenty-two-and-a-half minutes.

The rising and setting of the sun, of course, results from the spinning earth, with the globe rotating one complete revolution per day, i.e. 360° in twenty-four hours, or 15° per hour, or 1° every four minutes. The earth is also divided into 360° of longitude, 180° to the east and 180° to the west. At a position, therefore, time – which is basically measured by the sun – is related to the longitude of the position, with every 15° of longitude representing one hour. A point 45° east of the Greenwich Meridian, for example, is also three hours ahead of the time at Greenwich. Prochnow's time was ahead of Norfolk Island by twenty-two-and-a-half minutes which, with an earth rotation of 1° per four minutes, placed him 5.6° east of Norfolk. At about 30°S, 5.6° of longitude represented a range from Norfolk of about 290 nm. At a cruising speed of 110 knots, he could complete the journey in under three hours. Shortly before sunset his fuel remaining was three hours, perhaps more, so he might just be able to make it if only he could be found and pointed in the right direction.

With the sun below the horizon the lone pilot was once again enveloped in darkness. The Cessna droned on through the evening, heading roughly north-west in a desperate attempt to struggle closer to its destination. What Prochnow was feeling inside at that time can only be imagined for outwardly he remained cool. All his radio communications were calm and professional. It is a brave man indeed who can remain so composed when placed in such danger, for death was surely staring him in the face.

At the Auckland Search and Rescue Co-ordination Centre the co-ordinator, Bruce Millar, and his team were becoming increasingly concerned, so the RNZAF Orion on standby at Whenuapai was scrambled to join the search. The Orion would take about two-and-a-half hours to reach the Norfolk Island search area. Meanwhile, Prochnow continued with his series of radio calls on 121.5 MHz as the DC-10 sped away from his estimated position and out of range. As soon as communications were lost the position was plotted and 'Zulu Sierra' manoeuvred to sweep through the audio circle once again at right angles to the original traverse. It was almost an hour before the procedure was completed and the 'aural box' plotted on the chart. Allowing for movement of the Cessna, Prochnow's position was calculated at approximately 30°S 171°E.

'We estimate we'll be overhead your position in about five minutes,' radioed Vette to the Cessna. 'Turn in a circle and look out for our strobe lights.'

The strobes on 'Zulu Sierra' were powerful white flashing anti-collision lights which were so strong they could be seen clearly in daylight. If the Cessna was anywhere nearby surely the lights would be seen. Prochnow peered into the darkness, his hopes rising that at last he was to be found. The estimated time of rendezvous came, but once again the DC-10 was nowhere to be seen. With mounting despair amongst those involved, it had to be admitted that again an attempt at pin-pointing the Cessna's position had failed. Prochnow's spirits must have reached rock bottom. The DC-10 crew seemed to have exhausted all their options and now there was little they could do except sweep the area of the estimated position in the rather weak hope that Prochnow might spot their bright lights.

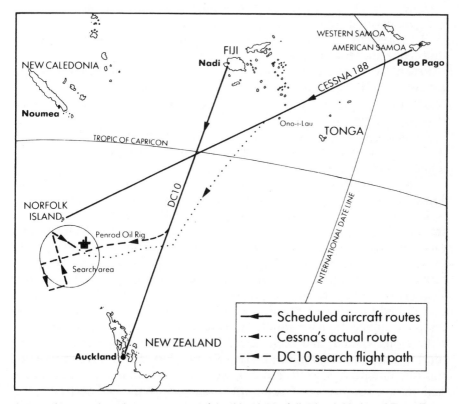

A map showing the relative positions of Auckland, Norfolk Island, Nadi and Pago Pago. The flight paths of the Cessna and DC10 are clearly marked.

The crew and passengers of the DC-10 peered desperately into the darkness but there was little chance of one of them spotting the dim lights of the Cessna in the blackness of the night. In a little over an hour the RNZAF Orion would be on station, and with its more sophisticated equipment it was hoped that the Cessna would be found quickly. Prochnow continued to fly north-west, hopefully reducing the gap between himself and Norfolk Island. If the Orion pin-pointed his position swiftly and he was given a precise heading to steer for the airport, he might just make it. Without help of some kind he was doomed, for he would never find the island on his own in such a vast stretch of sea. Prochnow and the DC-10 crew had functioned well as a team and they had all worked hard in the search. Having given of their best it seemed a gross injustice that the Cessna remained unfound. Prochnow had been in the air for twenty-and-a-half hours and he was exhausted with the flying and the concentration. The strain on him was enormous. What was needed now, more than anything, was luck.

'A light!' shouted Prochnow over the radio. 'I see a bright light on the surface. I don't care what it is, I'm going for it.' Any ship nearby would increase his chances of survival in a ditching.

'Aim straight for it and read me back your heading,' instructed Vette, 'but make sure it's not a star low on the horizon.'

Prochnow reported his heading as 310°. As the Cessna flew closer to the source of the light he was able to give a description of the vessel.

'It appears to be a tall structure riding on a platform,' informed Prochnow, 'there are two or three lights on it.'

'Sounds like an oil rig', said Dovey to the others on the DC-10 flight deck. Prochnow then reported seeing two tugs towing the vessel and all agreed it had to be an oil rig.

'Circle overhead and flash your lights to attract attention,' suggested Vette, 'we'll contact Auckland and obtain whatever information we can.'

The co-ordination centre was called on HF with a request for the marine division, part of the search organisation, to radio details of any oil rig platform being towed in the area. A few agonising minutes passed in waiting as the information was obtained, then back came the facts requested. It was identified as the *Penrod*, an oil rig being towed from New Zealand to Singapore, and its position was given as 31°S, 179°21'E. That established Prochnow's position as over 600 nm due east of Norfolk. He hadn't a chance. It also meant that the Cessna's position was hundreds of miles from where it had been placed by the navigators' estimates. Their methods had been rough and ready and only approximate fixes of position had been possible, but they were surprised that their calculations had been so far out. The Cessna was radioed with the bad news.

'You are too far east to make Norfolk on your available fuel,' informed Vette. 'It looks like you are going to have to ditch, but don't do anything till we get there. We can switch on our landing lights and brighten the area to help in your rescue when you go in.'

Dovey fed the co-ordinates of the *Penrod*'s position into the AINS and the big jet banked and flew towards the rig. Prochnow had already come to the same conclusion that ditching was inevitable, and in his mind was rehearsing procedures for the hundredth time. The rig had also spotted the Cessna circling overhead, flashing its lights, and the tugs had stopped their engines and hove to. It was obvious that the light aircraft was in trouble and a boat had been launched in anticipation of ditching. The *Penrod* was now fully illuminated and the rig was ablaze with lights.

'It looks like a giant Christmas tree', reported Prochnow.

In preparation for the ditching he lowered the Cessna's nose to inspect the sea.

'I'm going down now to take a look at the surface', he called.

Prochnow gently descended the light aircraft towards the ocean but what he found when he skimmed over the sea made his heart sink. There was a huge swell running, with waves fifteen to twenty feet (five to six m) high. In the wake of the current flowing past the rig's platform the surface was a little flatter, but hardly suitable for ditching. If the Cessna struck the top of a wave it could easily flip over and trap Prochnow in the sinking machine. If the little aircraft slammed into the side of a giant wave it would almost certainly break up on impact. Neither prospect was very appealing, yet Prochnow's best chance now was to put the Cessna down on the sea while he still had some fuel left. He could then use the engine power to control his rate of descent. If he landed as close to the *Penrod* as possible there was just a chance that the aircraft would remain upright and in one piece and that the rig's boat could rescue him.

'The surface looks bad,' reported Prochnow, 'I don't like it at all.' There was a touch of anxiety in his voice now but his approach was still very professional. Under the circumstances he remained remarkably cool.

The DC-10 sped on towards the rig's reported position but it soon became obvious that something was amiss. Instead of Prochnow's radio calls becoming clearer his transmissions were beginning to fade. 'Zulu Sierra' was flying away from the Cessna!

'There's something wrong with the rig's position,' radioed Vette to the search and rescue centre. 'Can you pass us a frequency so we can speak to him directly?'

Soon Vette was dialling the *Penrod*'s frequency of 119.1 MHz and contact was quickly established. Yes, they confirmed, a light aircraft was circling in the vicinity and they had launched a boat to rescue the pilot if he ditched.

'Can you confirm your position,' asked Vette, 'we have it as 31°S, 179°21'E.' Moments later the source of the problem was revealed; the centre in Auckland had mistakenly read a nine for a zero and one digit had been incorrectly transmitted. It was a simple mistake that anyone could have made, but one which altered the entire aspect of the scene. The correct position of the *Penrod* was 31°S, 170°21'E, and it didn't require the navigators to plot it on a chart to realise that the Cessna was much closer to Norfolk than the rig's incorrect position had first indicated. The new position of the *Penrod* was punched into the AINS, and as 'Zulu Sierra' turned back towards the target the Cessna was called with the good news.

'You may not have to ditch,' announced Vette, 'you are nearer Norfolk than we thought.'

The previous estimates of Prochnow's positions calculated by the navigators had, in fact, been remarkably accurate under the circumstances, and it had only been by the greatest of misfortunes that the two aircraft had failed to make visual contact and that Prochnow had been unable to spot the DC-10's powerful strobe lights. On retracing the steps on the chart it appeared that the fuel dumping had taken place directly over the top of the Cessna. It was possible that the fading light conditions had obscured the fuel trail from view. It was also very close to the aural box position adjusted for progress and the position line earlier indicated from Brisbane on HF.

A measurement by Forsyth from the chart revealed Prochnow to be only 150 nm from his destination, about one hour and twenty minutes' flying time away at the Cessna's cruising speed. Prochnow had now been in the air for about twenty one and a half hours, and only half an hour of fuel remained of his original endurance estimate. The Cessna had been flying at economy cruise for some time so as much as an hour's fuel could have been saved. That would give Prochnow the one and a half hours' endurance he needed for Norfolk. Whether to make a go for it or to stay put was Prochnow's decision alone, and not an easy choice to make. If the Cessna ran out of fuel on the way Prochnow would be in a worse situation than he was now. He would have to glide alone in the dark into the ocean swell with very little chance of survival.

The DC-10 flew at speed towards the *Penrod* with growing anxiety amongst the crew. It was now a desperate race against time.

'What's your estimate of remaining fuel?' enquired Vette.

'I've got about one quarter of one tank left,' suggested Prochnow, maybe about ten gallons (forty-five litres).' The Cessna's fuel consumption was eight gallons (thirty-six litres) an hour, giving a remaining endurance of about one and a quarter hours. It would be touch and go but he might just be able to make land.

The DC-10 had by now descended to 10,000 ft (3,050 m) to improve the chances of sighting the Cessna and already the bright lights of the *Penrod* rig could be seen ahead.

'We reckon you might just be able to make it,' announced Vette, 'what do you think?'

Prochnow took another look at the black and heaving sea and didn't need to be asked twice. 'I'll go for it', he said. 'Give me a heading and I'm off.'

'Steer 294° magnetic,' instructed Vette, 'we'll be with you shortly and we'll accompany you on the way.'

With a mixture of fear and hope in his heart Prochnow turned the little Cessna away from the friendly lights of the rig and back into the black night. F/O Dovey slowed the DC-10 to its minimum cruise speed of about 200 knots and turned slightly right to leave the *Penrod* on their left. Just then they spotted the little aircraft.

'The Cessna's been found!' announced Chief Purser James to the passengers. A great cheer rose from the cabin. While approaching the rig, the tiny Cessna could be seen from the flight deck, ahead and to the left, climbing away on its assigned heading. Prochnow was wisely gaining height to increase his gliding range in the event of the engine cutting with fuel starvation. The Cessna was difficult to see, even with its landing lights on, but this was a moment Captain Vette wished to share with his passengers.

'Ladies and gentlemen, if you look out of the left windows, about seventy degrees to the left and below, you'll see something very interesting. It's difficult to spot, but if you look carefully you'll see the dim lights of the Cessna.' Another cheer rose from the cabin. Eighty-eight pairs of eyes, with hands cupped at temples to shade the view, peered into the night for a sight of their quarry. It was a most satisfying moment for all.

The DC-10 sped past the tiny aircraft and Prochnow could clearly see the lights of the big jet above him and to his right.

'We'll set a course for Norfolk', suggested Vette. 'Follow our strobe lights, and if you lose sight let us know. We'll then double back if needs be and pick you up again.'

The DC-10 remained just sufficiently clear to prevent its jet wash striking the Cessna, and Prochnow tucked himself in behind and followed the airliner's path. Vette was also able to inform the light aircraft pilot that news had just been received of the RNZAF Orion's imminent arrival, and it would shepherd him home for the last miles to Norfolk. The drama was not over for Prochnow, however, for there was still a chance that the episode could end in disaster. With the passing of each mile the chances of reaching his destination improved, but with the consumption of each unit of gasoline so did the risk of running out of fuel. It was a most anxious period of the flight. Prochnow had now been in the air for twenty-two hours and had reached the extent of his original estimated fuel endurance. He was now living on borrowed time and saved fuel, and at 100 nm from

Norfolk he still had one hour to go. Prochnow held on as best he could and steadfastly followed the bright strobe lights of the DC-10 which could still be seen ahead.

'You are going to make it,' encouraged Vette, 'just be sure you use all of your available fuel. Check every tank in turn and run each one dry.'

The RNZAF had caught up with the search by now, and as the Cessna pilot looked to his left he saw the Orion formate off his left wingtip. It was still going to be a close-run race, but at least help was at hand.

At fifty nautical miles to Norfolk, 'Zulu Sierra's' strobe lights remained in Prochnow's sight as the DC-10 turned over the island. The Cessna hugged the side of the Orion. If a last minute ditching resulted, the Air Force crew were ready to aid his recovery. There was little more now that Vette could do and, with the DC-10 crew's tasks completed, the big jet turned for Auckland while the Orion oversaw the homecoming of their charge. Anxiously they listened on the radio for news. At last the Cessna, accompanied by the Orion, came into view at Norfolk, and after twenty-three hours in the air Prochnow approached the airport and turned onto short finals. At least now he was over dry land. Five minutes later, with virtually dry tanks, he touched down on the runway at the local Norfolk time of midnight, a relieved and happy man to be back on solid ground. The Cessna was eight hours late on arrival and twenty-two hours of fuel had been stretched to twenty-three hours and five minutes.

'The Cessna has landed,' announced Captain Vette to his passengers, 'he is safe at last.' The apology which followed for the delay was drowned by cheering and clapping from the cabin.

Vette and his crew, however, still had to complete the work on their own flight so, with the search satisfactorily concluded, they settled down for the one-hour journey remaining to Auckland. As the airliner sped on course, Prochnow called 'Zulu Sierra' on the radio while taxiing in at Norfolk and expressed his sincere thanks for the DC-10 team's effort and for their work in helping to save his life.

'I think it's time to celebrate,' said Vette to his Chief Purser, 'break open the champagne.' In the cabin, eighty-eight glasses were raised in salute to the gallant rescuers.

The entire search and rescue exercise had been an outstanding episode. The two navigators, Captain Vette and F/O Forsyth, had worked ceaselessly at their task for over three hours, while F/O Dovey and F/E Brooks had operated the aircraft over the period at a continuously high workload. It was an admirable achievement. Captain Vette remarked of his team, 'The fact that they were such exceptional airmen helped a great deal.' Prochnow's efforts were not to be forgotten either, for his calmness and professionalism throughout were a credit to the piloting fraternity.

At 01:29 local New Zealand time the following morning, Flight TE 103 arrived at the gate at Auckland international airport, three hours and fifty-four minutes behind schedule. For once, on a delayed service, no-one complained.

The Guild of Air Pilots and Air Navigators awarded Gordon Vette and Malcolm Forsyth the Johnstone Memorial Trophy in 1980 for outstanding air navigation. McDonnell Douglas, the DC-10's manufacturers, also awarded the two navigators, as well as Arthur Dovey and Gordon Brooks*, a certificate of commendation for displaying 'the highest standards of compassion, judgement and airmanship'.

* Gordon Brooks was killed in the crash at Mount Erebus, Antarctica, in 1979. Gordon Vette went on to write a book of the accident entitled *Impact Erebus*, published in 1983.

Chapter 4

The Windsor Incident

'Shall we give it a try?' asked Captain McCormick.

'OK, I'll switch off the hydraulics.'

The instructor turned to the flight engineer's panel and selected the hydraulic pumps to off.

'You're on your own now.'

With hydraulic pressure reduced to zero the flying controls became ineffective and McCormick was left with engine power only to guide the machine. Both men watched the performance with interest and were pleased with the result. The DC-10 'flew' quite well and could be controlled with reasonable success on engine power alone. A bit of practice was all that was required.

Captain Bryce McCormick was a very experienced pilot with 24,000 flying hours to his credit. In the spring of 1972 he was fifty-two years of age and had been with American Airlines since before the end of the war. In his career he had flown a number of aircraft: Convair 240, DC-3, DC-4, DC-6, DC-7 and Boeing 707. In early 1972 McCormick had converted to the DC-10 and by the end of March had completed his ground school, simulator and flying training. In April he recommenced line flying duties on his new aircraft type.

McCormick liked the DC-10 and thought it a fine machine, but his admiration was not entirely without misgivings. The DC-10, like the Lockheed Tristar and Boeing 747, was one of the new generation of 'wide-bodied' aircraft which were introduced into service in the early 1970s. Not only were these aircraft bigger with more powerful engines and larger passenger capacity, but they also incorporated new design concepts. Conservative pilots like McCormick, steeped in traditional values, felt uneasy about accepting some of the more radical changes. All jet transport aircraft, for example, have hydraulically operated flying controls, but the three 'wide bodies' had no manual back-up facilities to operate the controls in the event of complete hydraulic failure, unlike the earlier-generation Boeing 707 which McCormick had flown.

Back-up systems normally consisted of cables running from the control column which were connected directly to the elevators, ailerons and rudders. With total hydraulic power loss the pilot could, albeit with some difficulty and a lot of muscle, successfully retain control of the aircraft.

Captain Bryce McCormick.

The 'wide bodies' also employed control cables, but these only fed demands to hydraulic power control units which in turn deflected the control surfaces. Any pilot of one of the newly introduced 'big jets', faced with the unlikely event of total hydraulic power loss, would have no flying controls available to fly the aircraft. What would happen in these circumstances? Could the aircraft be turned, climbed and descended by simply varying the power of the appropriate engines? It was a question that McCormick had pondered on a number of occasions during his DC-10 conversion course and one that he was determined to find out. He now sat in an American Airlines DC-10 flight simulator at their training school at Fort Worth, Texas, and, with the aid of an instructor friend, was experimenting in guiding the aircraft on engines alone.

'It handles pretty well', commented McCormick.

Both men were pleased with the outcome but not entirely surprised by the success of the exercise. The DC-10, in particular amongst the three 'big jets', has its engines exceptionally well placed for steering by engine power only: one on each side slung in pods below the wings about one

third of the way out from the wing roots with the lines of thrust sufficiently far apart for directional control and also directed below the
fuselage, and one placed high on the lower half of the tail fin with its line
of thrust directed above the fuselage. The Tristar is similar, but the
DC-10's highly placed tail engine makes pitch control easier. By using
asymmetric power, i.e. by increasing power on one wing engine and
decreasing the other, the aircraft could be turned, not unlike the differential steering of a tank by varying the speeds of the tracks. Likewise, the
aircraft could be climbed by pitching the nose up using increased power
from the low-slung wing engines and decreased power of the highly placed
tail engine, and vice versa for descent. McCormick spent thirty to forty-
five minutes practising the technique and became quite adept at 'flying'
the DC-10 simulator in such a manner. By the end of the exercise he could
control the simulator from initial climb to approach phase using only the
thrust levers without touching the flying control column. McCormick was
satisfied with the outcome of the experiment and his success in handling
the stricken machine helped allay some of his concern. He was not
to realise at that time, however, that in only a few short months his
expertise would be put to the test in earnest.

On the morning of 12 June 1972, American Airlines Flight 96 took off
from Los Angeles, California, on a scheduled service to New York's La
Guardia Airport, routeing via Detroit Metro, Michigan, and Buffalo in
the western tip of New York State. Flight 96, a DC-10-10, registration
N103AA, was commanded by Captain McCormick, with First Officer R.
Paige Whitney as his co-pilot, and Flight Engineer Clayton Burke. In the
cabin eight stewardesses were led by Chief Flight Attendant Cydya Smith.
The DC-10 departed forty-six minutes late from the American West
Coast because of passenger-handling difficulties and air traffic control
delays. After an uneventful four-hour flight, *N103AA* arrived in Detroit
at 18:36 Eastern Standard Time. Of the few passengers aboard, thirty-
eight disembarked, while sixteen remained, to be joined by forty others
for the next leg. Captain McCormick hoped for a good turn-round time
to help make up some of the delay. With only fifty-six passengers travelling on the next sector and little fuel required for the short forty-five
minute hop, there was every likelihood of a swift departure.

Soon ground preparations were completed and the passenger doors
closed. The two main cargo doors on the right side, situated in the mid-
and-forward positions, were checked closed, but the smaller aft bulk
cargo hold door on the left side was causing difficulty. It was not the first
time that problems had occurred in the locking of this door, and all four
airlines operating the DC-10 at that time (American, Continental,
National and United) had experienced trouble. The door was shut by an
electric motor and actuator and was then secured in position by latches
being wound on to spools (see diagram). The door handle operated a

An American Airlines DC-10. *(John Stroud)*

locking mechanism and when placed flush with the door slid locking pins into place to prevent movement of the latches. Closing of the handle also shut a small vent door and extinguished the 'door open' light on the flight deck. In such a condition the door was safe and would not burst open in flight under the force of the air in the pressurised cabin.

The problem being encountered with the aft cargo door was that the actuator being powered by the electric motor was not driving the latches fully home and it was necessary on a number of occasions to use a hand crank to complete closure. McDonnell-Douglas were already aware of the circumstances and had issued service bulletins recommending rewiring of the electric motors with heavier-gauge wire. Captain McCormick's aircraft, unfortunately, had not yet been modified. On the transit in Detroit the ramp service agent closed the door electrically and listened for the motors to stop running. Cut-out of the electric motor should have indicated the correct positioning of the latching mechanism, but once again problems occurred. The latches were not driven fully home and the latch linkages were not moved to the over-centre position (see diagram). Flanges obstructed movement of the locking pins and prevented the door handle from being placed flush with the surface. Using a certain amount of pressure the ramp agent could feel movement of the handle and simply assumed that the mechanism was stiff. Eventually he managed to force the handle into position using his knee and with the lever properly stowed assumed the locking procedure to be completed.

The latches, however, were still improperly seated and the door remained unlocked. With the locking pins jammed against the restraining flanges the forced closure of the door handle had buckled the locking rods. The linkage had moved sufficiently, however, to close the vent door, although in a slightly cocked position, and to extinguish the 'door open' light on the flight deck. Not surprisingly the ramp agent was unhappy with the circumstances, and he called over a mechanic to examine the vent door, which could be viewed through the vent aperture. The mechanic could see the vent door slightly out of position but all the indications appeared normal. The state of the door, however, was anything but normal and was in a most unsafe position. In the cockpit the flight engineer checked the 'door open' light extinguished and could only assume that the door was closed and locked. Captain McCormick and his crew were unaware of any danger.

On Flight 96's departure from Detroit, compressed air would be tapped from the engine compressors to pressurise the cabin for relatively normal breathing for those on board as the aircraft climbed into the thinner air. The aircraft hull and doors were designed to contain the pressurised air in the cabin from bursting outwards into the rarefied atmosphere. To help ease the exertion on the structure, however, cabin pressure is reduced as

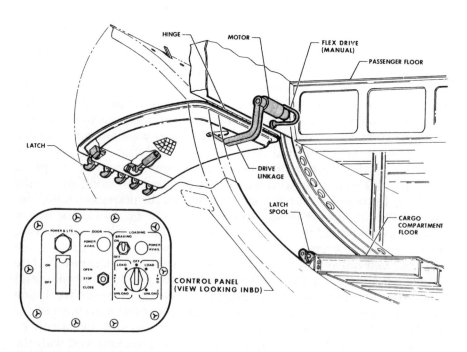

A DC-10 aft cargo door operating mechanism. *(NTSB)*

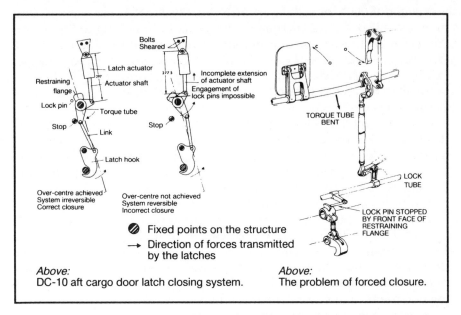

Above:
DC-10 aft cargo door latch closing system.

Above:
The problem of forced closure.

The aft cargo door latch closing system – the problem of forced closure. *(NTSB)*

the aircraft climbs (for example, to an equivalent cabin altitude of 6,000 ft (1,830 m) when cruising at 35,000 ft (10,670 m), but the outwards force is still quite considerable. Would the partially locked aft cargo hold door be able to withstand the strain of several tons of pressure as the cabin pressurised after take-off? If not, and the latches released under the pressure, the door would blow off. The rapid venting of the pressurised air to the atmosphere would cause untold damage and could place the aircraft in great jeopardy.

Meanwhile, unaware of the insidious danger of the situation, the DC-10 crew continued with their departure preparations. As the mechanic inspected the aft cargo door, Whitney called on the radio for departure instructions. The time was just after 19:03.

'Metro clearance delivery from American 96, information "delta", and airways to Buffalo.'

'Delta' was the current weather details transmitted by the automatic terminal information service and indicated the cloud base to be 4,500 ft (1,370 m), the visibility reduced to 1–1½ miles (1.6–2.4 kilometres) in smoke and fog, a south-west wind at six knots, a temperature of 61 °F and an altimeter pressure setting of 29.85 inches of mercury. The weather, although not good, would give no problems on departure. Clearance delivery replied instructing American 96 to turn right after take-off onto a heading of 060° for radar vectors to airway Jet 554. The flight was to maintain 4,000 ft (1,220 m) and could expect to be cleared to flight level 250 (25,000 ft, 7,620 m) ten minutes after departure. The departure frequency was given as 118.95 MHz and the squawk, or radar identification code, as 0200. Whitney repeated the clearance as read, but the controller corrected himself and reissued a departure frequency of 118.4 MHz.

In the cockpit the crew completed their pre-start checks. A few minutes' delay ensued as the aft cargo door inspection continued, but finally the door condition was accepted and the all-clear was received on the intercom from the ground engineer. The engine start sequence was commenced and Whitney contacted ground control on 121.9 MHz.

'Ground, American 96, push from gate ten.'

Clearance to push was given, and as the push-back truck revved its engines the DC-10 moved slowly backwards from the gate. The time was 19:11 Eastern Standard Time and the turn-round had been an excellent thirty-five minutes. McCormick was pleased to be making up some time. A few minutes later, with all engines running, Whitney requested permission to taxi and ground quickly replied.

'American 96, taxi to runway three right.'

On the taxi out the take-off performance on runway 03R was checked and the 'before take-off' check list commenced. Approaching the threshold Whitney called the tower on 121.1 MHz.

'American 96,' replied the tower controller, 'into position and hold.'

The DC-10 lined up on the runway and with the anti-skid selected the flight engineer pronounced the checks completed. First Officer Whitney was to fly the aircraft on this sector and he stood-by for the take-off. The captain took over operation of the radio.

'American 96, maintain runway heading and contact departure. Cleared for take-off.'

Whitney opened up the power to the required setting and the lightly laden big jet rapidly accelerated down the runway. At 19:20 the wheels lifted from the surface and *N103AA* climbed easily into the air. McCormick established contact with the departure controller.

'American 96, Detroit departure, radar contact. Climb and maintain six thousand.'

'OK,' replied McCormick, 'you want us to climb and maintain six thousand, American 96.'

'Roger, American 96, now turn right heading zero six zero.'

Whitney increased the DC-10's speed and the flaps were selected from the take-off position to up.

'American 96, turn right heading zero nine zero.'

Flight 96 was being carefully radar vectored to pick up the easterly airway Jet 554 en route to Buffalo.

'American 96, turn right heading one one zero. Join Jet five five four when you intercept.'

The DC-10 now entered cloud, and as the aircraft climbed rapidly towards its cleared altitude of 6,000 ft (1,830 m) McCormick replied asking if they were to maintain this height.

'OK, I'll have something higher for ya in just a moment,' replied departure, 'climb and maintain, ah, flight level two one zero.'

Clearance for further climb arrived just in time for the '1,000 to go' alert tone sounded in the cockpit.

'OK, ah, climb two one zero, American 96,' repeated McCormick, 'we are out of, ah, fifty five hundred now.'

'American 96, call Cleveland Centre now on frequency one two six point four. Good day.'

The US/Canadian border lies lengthwise along the centre of Lake Erie and most of Flight 96's journey would be spent in Canadian airspace. Since the DC-10 was operating between two American airports, however, US controllers continued to direct the flight. *N103AA* was already over Canadian territory and was now close to the town of Windsor in Ontario.

In the cockpit McCormick quickly selected the Carleton radio beacon on both navigation receivers and set the course indicators to 083°, the Jet 554 airway centre line, to permit interception of the airway. As the DC-10 settled on the heading of 110° Whitney engaged the autopilot, but out of habit kept his hands on the control column. The Cleveland frequency was dialled on the radio box and McCormick called the Centre.

'Good evening, Cleveland Centre, American 96 is out of, ah, seven thousand now for two one zero.'

'American 96, squawk code, ah, one one zero zero and ident. Maintain flight level two three zero, report reaching.'

The DC-10 continued towards Jet 554 and, still in cloud, climbed at the maximum permitted speed below 10,000 ft (3,050 m) of 250 knots to its newly assigned flight level. At 19:24, only four minutes after lift-off, the lightly laden aircraft passed through 10,000 ft. Whitney lowered the nose of the DC-10 using the autopilot vertical speed control to accelerate the machine to the normal climb speed of 340 knots. The aircraft's rate of climb reduced to the 1,000 ft (305 m) per minute selected and the speed began to increase. As the flight approached the top of the weather the cloud thinned and above could be glimpsed the first signs of the sun. A few moments later, passing 11,500 ft (3,500 m), the DC-10 broke from the cloud layer into a bright mid-summer's evening. 'What a beautiful day', thought Whitney to himself.

N103AA passed over the town of Windsor, situated by the shore of Lake St Clair lying over two miles (3.2 kilometres) below, and continued towards Jet 554 as it headed east-south-east between the Great Lakes of Huron to the north and Erie to the south. At only four-and-a-half minutes after take-off the DC-10 climbed through 11,750 ft (3,580 m) and the increasing speed edged past 260 knots. Captain McCormick looked out into the clear sky and could see far above a giant Boeing 747 flying majestically through the air.

'There goes a big one up there.'

Whitney leaned forward to catch a better view of the other aircraft, still resting his hand gently on the control column. It was a pleasant moment in the trip with the weather and the busy departure behind and a clear flight ahead, but it was a calm which did not last long. Suddenly an enormous 'thud' was heard from the rear of the aircraft and in an instant their peace was shattered. The air 'fogged' and a great rush of air swept past the flight crew. Dirt and dust flew in their faces and stung their eyes and skin as if a firecracker had gone off below their noses. For a moment they were blinded.

'Oh . . . !' someone shouted.

The rudder pedals exploded and smacked at great speed to the full left rudder position. The captain was resting his feet on the pedals and his right leg was thrown back against the seat with extreme force. His right knee was driven to his chest and his headset was knocked from his ears. First Officer Whitney, still leaning forward, was thrown violently rearwards and hit his head as he crashed against his seat. At the same moment

The flight crew: Captain Bryce McCormick (left), First Officer R. Paige Whitney (centre) and Flight Engineer Clayton Burke (right). *(Captain Bryce McCormick)*

the three thrust levers snapped to flight idle with the number two (tail) engine throttle hitting the stop with a loud crack. The aircraft was felt to 'jerk' momentarily and the autopilot disengaged with the red disconnect light flashing. McCormick feared a mid-air collision had occurred and suspected the windshield had shattered. When his eyes cleared he could see it was still in place but, disbelievingly, he stretched out his hand to touch the window.

'What the hell was it, I wonder?' he called.

One of the crew replied with a long whistle. Captain McCormick now noticed a red failure flag on his airspeed indicator and speculated that the radar dome might have blown off. With the aircraft nose cover missing erratic speed indications could be expected. Whitney still maintained his hands on the column but the DC-10 yawed and banked slowly to the right with the nose dropping sharply out of control.

'Let me have it', yelled the captain as he grabbed the control wheel. Quickly he tried to 'feel out' the controls.

'I think it's going to fly', reassured Whitney.

At three seconds after the initial explosion an engine fire warning sounded, together with the *beep, beep, beep* of the cabin altitude warning horn. The DC-10 appeared to have suffered an explosive decompression and the intermittent horn indicated that the cabin air pressure had reduced to an equivalent altitude in excess of 10,000 ft (3,050 m). McCormick had no doubt that he had a 'pretty sick airplane' on his hands and the first priority was to keep it in the air. 'Fly the airplane, fly the airplane', he kept saying to himself. The engine and pressurisation problems could wait.

'We'll pass the fire warning', he called.

Captain McCormick was an advocate of the theory that he who hesitates survives. When faced with a sudden emergency he firmly believed that a pilot's first action should be to do nothing – think the problem out, then act. In normal circumstances, with an intermittent horn warning of reduced cabin pressure, the crew would commence an emergency descent, but the captain was reluctant to force the aircraft into a steep dive until the damage could be assessed. If he did drop down quickly he might not be able to recover. Flying at around 12,000 ft (3,660 m) few breathing problems would be experienced, although strictly speaking flight crew should don oxygen masks above 10,000 ft. The engine fire, too, was a serious problem, but a more pressing matter was the need to keep the aircraft flying. There would be little point in extinguishing a fire if the DC-10 tumbled out of control. To the crew their actions felt strange, like moving in slow motion.

'We've hit something', said the flight engineer.

The co-pilot had been looking out at the time and had not seen anything of another aircraft so thought it more likely to be disintegration of number two engine. That would explain the fire warning and the rudder problem.

'We've lost . . . lost an engine here', he said.

'Ah, which one is it?' asked the captain.

Flight Engineer Burke could see the problem from the engine instruments.

'Two. Number one is still good . . . and, ah, Captain . . . we'll have to . . . to check this out.'

McCormick could see the aircraft descending towards the cloud tops which lay below at 10,000 ft and he wanted to avoid penetrating the weather until he was sure he had control of the stricken machine. Steadily he heaved back on the control column but the response was very sluggish. The speed had dropped to 220 knots and simultaneously he pushed the three thrust levers forward. The numbers one and three (wing) engines responded but the number two (tail) engine remained in flight idle. The number two throttle could be moved backwards and forwards quite easily and it obviously wasn't attached to anything. Immediately the DC-10 responded to the power increase from the wing engines and the nose pitched up. McCormick only just managed to stay in the clear.

'OK,' said Whitney, 'we apparently . . . master warning . . . this board's got an engine fire over here. Yeah, we got two engines, one and three. Do we have, ah, hydraulics?'

'No,' replied McCormick, 'I've got full rudder here.'

The rudder was still jammed with the left pedal fully forward and the captain assumed the condition to be the result of hydraulic failure. The flight engineer checked out the systems, however, and found them satisfactory.

'Hydraulic pressure is OK', he said.

In the cabin there was just as much confusion as there was in the cockpit, and some of the passengers were hysterical. Chief Flight Attendant Smith rushed forward to the flight deck to check with the captain.

'Is everything all right up here?'

'No', called McCormick. The co-pilot turned and shook his head. 'You go back to the cabin.'

The DC-10's speed established at 250 knots and Captain McCormick managed, with difficulty, to hold the aircraft clear of cloud at 12,000 ft. While the other two continued to check out the systems he retrieved his headset from the back of the seat and quickly radioed Cleveland Centre declaring an emergency. There was little information he could give as they didn't know what had happened except that they had a serious problem.

'Ah, Centre, this is American Airlines Flight 96, we got an emergency.'

'American 96, roger', replied Cleveland. 'Returning back to Metro?'

'Ah, negative, I want to get in to an airport that's in the open. Where's one open?'

Having just left Detroit Metro with a visibility of one to one-and-a-half miles in smoke and fog and having climbed through a 6,000 ft thick cloud layer, McCormick was reluctant to return there until he was sure of the integrity of the flight controls. It would be more sensible to avoid flying on instruments if there were control problems and to head somewhere clear of weather. He would make a decision once he had checked out the DC-10's controllability. The aircraft was continuing to fly, albeit with difficulty, but there was no doubt that the controls were severely impaired. The rudder pedals were solidly jammed, with the left pedal fully forward, but with the rudder itself trailing slightly to the right. The effect yawed the aircraft nose to the right and the aircraft flew in an askew condition. The elevators controlling climb and descent were extremely stiff to move and produced a very sluggish response.

McCormick tried to trim out the elevator control column forces by operating the electric stabiliser trim switches, but to no avail. Trimming consists of varying the angle of the tailplane to the airflow to stabilise an aircraft in flight, and as such the variable tailplane is known as the stabiliser. On McCormick's aircraft the stabiliser position indicator remained fixed and the stabiliser did not function. The captain then tried to operate the manual trim control but as he did so the handle came away in his hand. So much for that! There was little McCormick could do now to ease the elevator strain but he felt he could overcome some of the difficulty in pitch control by varying the power of the wing engines. The offset rudder kept rolling the aircraft to the right, but, with the control wheel turned 45° to the left, it was possible to use the ailerons, which are normally employed to bank the aircraft for a turn, to hold the wings level. In this condition, turning of the aircraft, especially to the right, was a delicate operation and only slow 15° banked turns could be achieved.

In a few moments, by experimenting with the controls, McCormick quickly brought the stricken airliner under control. He sensed now that he could keep the aircraft in the air, but could he land it in such a condition? That was going to be a different problem. Flying the DC-10 on engines was one thing, but guiding a large jet in such a manner onto a 200 ft (sixty m) wide runway was a different matter. Unfortunately there was little choice but to try.

McCormick's thoughts now turned to his passengers, who he suspected would be in some distress. Quickly he picked up the public address (PA) handset to offer reassurance. He had little to tell them but he felt any words of encouragement would be better than none. Trying to hide his own apprehension he spoke as calmly as possible saying that they had a mechanical problem and that they would be returning to land.

'American 96,' called Cleveland Centre, 'start right turn. Heading'll be one seven zero, maintain ten thousand.'

As the stricken aircraft banked slowly to the right, the flight crew pondered the problems which lay ahead. As yet no one had any idea of what had happened. Back in the cabin Chief Flight Attendant Cydya Smith took stock of the situation and began assessing the damage. She knew that the captain would expect a full report as soon as possible. As she proceeded rearwards, comforting and assuring passengers, she was amazed at the destruction she could see towards the rear. At the time of the 'explosion' she had been standing in the service centre preparing coffee and drinks for the passengers. The door of the lift to the lower galley suddenly burst open, narrowly missing her head, and out billowed what appeared to be a 'smoky substance'. She was thrown off balance but managed to steady herself by grabbing a handrail. She felt a sensation of weightlessness which was followed by complete silence. Instantly she thought the aircraft had suffered a depressurisation and quickly checked to see if oxygen masks had dropped in the cabin. They remained in position and she had no difficulty in breathing. (At cabin altitudes above 10,000 ft (3,050 m) the warning horn on the flight deck sounds but it is only at cabin altitudes above 14,000 ft (4,270 m) that passenger oxygen masks drop automatically.) It was then that she rushed forward to the flight deck to check on the situation and on the way had been puzzled to see the captain's hat lying on the floor of the front cabin. In the cockpit she had seen from the expressionless look on the co-pilot's face as he spoke to her that something was seriously wrong so she had returned immediately to the cabin. She then spoke on the PA, instructing the travellers to stay in their seats, not to smoke and to remain calm.

Suddenly Stewardess Carol Stevens grabbed her arm and told her that one of the flight attendants at the back was trapped by the edge of a large hole and that she urgently needed help. Stevens had been sitting with her seat belt still fastened when she heard a 'varoom' type of noise and the cabin had fogged. Air had rushed from front to rear of the aircraft. When the mist cleared she had seen the devastation in the rear section of the cabin. A small lounge bar was situated in this area for the use of economy passengers, but since few people were on board and the sector was short it was considered not worth offering the facility. No one was seated in the area, which was just as well, for the floor had caved in and ceiling panels were hanging down. The bar unit had collapsed and had fallen into the hole on the left side of the aircraft. On the floor and halfway into the cavity lay Stewardess Bea Copeland, prostrate on her side but facing forwards with her head towards the centre. She was trapped by the wreckage and was shouting for help. Stevens had tried to call Cydya Smith to tell her of the girl's predicament but the cabin interphone failed to work. She was hesitant to leave her seat since she assumed the cabin had decompressed and she was expecting an emergency descent. Eventually she noticed some of the other attendants out of their seats so she had run forward to raise the alarm.

On take-off Bea Copeland had been positioned by the aft left exit and had remained in her seat with her seat belt fastened. There were few passengers on board and, with time to spare, she had sat chatting to Stewardess Sandra Mcconnell who sat by the aft right exit. The bar area was situated immediately forward of the girls' positions. Copeland had only just released her seat belt when suddenly she had been lifted from her seat and thrown to the floor. A ceiling panel had detached and had fallen on her head and debris had trapped her foot. From where she lay she could see down into the cargo hold. The large bar unit had crashed behind her, partially blocking the hole and preventing her from falling further. She had tried to shift the panel covering her head but was unable to move it so she had begun calling for help. On the right side Stewardess McConnell had been thrown from her seat with the 'explosion' and had struck the divider position behind the bar. She had landed on the damaged floor and had felt the area around her begin to crumble. From where she lay she could see daylight through the side of the aircraft and she could feel herself slipping slowly into the hole.

Chief Flight Attendant Smith, with Carol Stevens and a male passenger who offered to help, rushed back to assist the trapped stewardesses. One of the other attendants, unaware of the rescue in progress, called the flight deck on the interphone to ask if someone could lend a hand. In the cockpit the flight crew had their own problems, of course, but they seemed to be overcoming the initial difficulties. As Cleveland Radar vectored *N103AA* southwards for a return to land, the crew quickly rearranged their duties. Until the moment of the incident the co-pilot had been flying the aircraft. The captain had taken control but had continued to operate the radio while the other two inspected the systems. They had checked out the number two engine fire warning and found it to be false. McCormick continued to fly the aircraft while Whitney now took over communications.

'We've got one seven zero heading, sir,' called the co-pilot to Cleveland Control, 'and, ah, maintaining twelve thousand.'

'96, roger, type of emergency?'

Whitney misheard and simply replied, 'Yeah, yes sir.'

'We have a control problem,' McCormick added, 'we have no rudder, got a full jam. We've had something happen, I don't know what it is.'

'American 96, understand . . . cleared to maintain, ah, niner thousand, altimeter two nine eight seven. Be, ah, radar vector back towards the ILS course runway three. You want the equipment standing by?'

Cleveland was suggesting the DC-10 return to Detroit Metro, landing

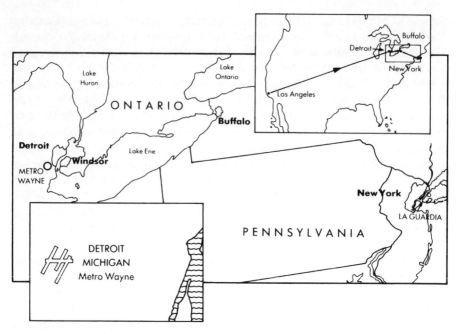

Map showing the relative locations of Detroit, Windsor, Buffalo and New York. Inset shows the DC10 flight path from Los Angeles.

in the same direction as take-off with radar guidance onto the instrument landing system (ILS) for either of the parallel 03 runways.

'OK, sir,' replied Whitney, 'ah, say again the heading and we'll let down slowly to nine thousand.'

'Heading'll be two zero zero . . . two zero zero for American 96. And do you wish the equipment to be standing by?'

'Affirmative.'

'Understand. Full equipment standing by at Metro.'

'Yes, sir.'

The interphone now chimed in the cockpit and Flight Engineer Burke answered. It was the call requesting help for the trapped stewardess. The floor had collapsed, explained the caller, and she was lying by the edge and in need of assistance. Clayton Burke discussed the situation with the captain and since the immediate flying problems seemed to be resolved the flight engineer left his seat to help. Aware of the need to be properly dressed in the cabin he searched for his hat, which was missing from the clips on the flight deck door. He looked in the crew wardrobe but it was nowhere to be seen. Unknown to Burke the initial rush of air had blown the crew's hats from the back of the door as it flew open and they lay scattered on the cabin floor. Stewardess Carol McGhee had been sitting

72

strapped in by the front entrance at the time of the incident and had seen the flight deck door burst open and the escape hatch from the downstairs galley shoot up and strike a passenger on the head. A dusty rush of air had gushed from the flight deck, together with the flight crew's hats, which had flown out at head height as if worn by ghosts bolting from the cockpit. Hence the captain's hat seen by Cydya Smith lying on the floor and the flight engineer's inability to find his own.

At the back of the aircraft Bea Copeland still struggled on her own to free herself from the wreckage. Several times she tried to move the panel covering her head and eventually she managed to push it away. Her shoe was stuck in the debris and she was unable to move it, but with some effort she managed to pull her foot out of the shoe and set herself free. She then began to climb out of the area and as she struggled over the bulkhead was met by Smith and the others. Her rescuers helped her onto the rear lounge seats and then grabbed her hands and pulled her into the economy section. They now looked for McConnell who had been sitting by the aft right exit but they could see no sign of her and feared she had fallen into the hold. Smith called out her name repeatedly, but to no avail.

'Sandy, Sandy, are you OK?'

Suddenly as if from nowhere, she appeared in view. She had managed to pull herself out of the hole and, somewhat dazed, had clambered into a rear toilet. The floor appeared very weak and she had felt that it would have collapsed if she had walked over it. Fortunately those on the scene were now able to help and with some difficulty manoeuvred her over the rear seats and into the cabin area. She was badly shaken, as was Copeland who had been trapped on the other side, but otherwise both girls were unhurt. It had only been a few seconds before that the call for help had been put through to the flight deck so one of the girls now rang to say that the two flight attendants had been rescued. Flight Engineer Burke had only just started to look for his hat when the good news was received, so he simply resumed his seat.

At the back of the aircraft Chief Flight Attendant Smith instructed the stewardesses to move away from the damaged area and to shift all the passengers nearby to forward sections. She then returned to the flight deck to report to the captain. Smith informed McCormick of the destruction at the rear of the cabin and that 'there was a hole in the fuselage in the very aft of the left-hand side'. The situation was worse than any of them had expected and they had done well to keep the aircraft flying. The reason for the DC-10's plight was still a mystery, for the problem with the locking of the left rear cargo door was unknown to the crew.

As the fuselage interior had pressurised on the climb out from Detroit the partially locked latches had been subjected to an increasing force. About five minutes after take-off, as Flight 96 passed through 11,700 ft (3,570 m), the pressure on the door was in excess of five tons. The door's electric actuator bolts had suddenly sheared under the stress and the door latches had sprung. The door had blown open, causing explosive decompression of the aircraft. The door had been torn off by the airstream, damaging the left tailplane in the process. The pressurised cargo air had immediately exhausted to the atmosphere via the gaping hole and the cabin air pressure had placed an undue load on the floor. With insufficient venting in the cabin floor area the floor had simply collapsed, tumbling the bar unit into the gap.

Unfortunately the structural devastation was not the only damage sustained. Through the beams of the cabin floor ran a number of vital control cables, hydraulic pipes, fuel lines and wiring to the empennage control surfaces and the number two engine in the tail fin. Many had been either severed or jammed, or had their operation severely curtailed. Ironically, McCormick's original fears of total hydraulic failure remained unfounded in this circumstance for the three hydraulic system lines, like the fuel line to number two engine, remained intact. They did not run through the collapsed central section of the floor and, with the lines coiled at various points to allow for stretching, had survived the impact of the decompression.

The flying controls at the tail, however, functioned via power control units which, like the number two engine controls, were operated by cables which ran from the flight deck through the central floor section along the entire length of the aircraft. As the pilots manipulated the control columns, or engine thrust levers and fuel controls, the taut and finely adjusted control cables moved in response. When the rear cabin floor collapsed the left rudder cable was broken, allowing the right rudder cable to slacken. The intact right cable was then forced downwards by the fallen floor which had deflected the rudder to the right, jamming it in that position. The taut cable, however, had pulled the right rudder pedal rearwards, giving an opposite indication on the flight deck. Three of the elevator control cables were severed and only the right outboard elevator panel remained functioning. The downwards load of the collapsed floor also resulted in the elevator being extremely difficult to operate. The tail engine thrust lever cable and the fuel shut-off valve cable were also broken. Severed wiring had resulted in the spurious number two engine fire warning.

McCormick informed his chief flight attendant that they expected to be landing in about eight to ten minutes and he instructed her to prepare for an emergency landing. Smith enquired if they'd be evacuating via the chutes, but since the captain was uncertain of what would happen when they touched down he was unable to say. As the discussion continued Cleveland Centre called again.

'American 96, you say you believe you hit something?'

'Ah, I don't know, sir', replied Whitney. 'Just standby one. Standby one until we assess the situation. We've got here, ah, definite problems.'

McCormick interrupted the exchange.

'OK, now we have got, ah, problems. I got a hole in the cabin, I think we've lost number two engine, we got a jammed full left rudder and we need to, ah, get down and make an approach. I guess Detroit Metro would be best and, ah, can you vector us around?'

McCormick felt the DC-10 to be flying satisfactorily, so Control's earlier suggestion of landing back at Detroit seemed a reasonable course of action.

'American 96, roger', replied Cleveland. 'Turn further right now, heading'll be two zero zero.'

'OK, two zero zero, American 96.'

'Roger. And that the left, ah, rudder is jammed?'

'Er, it's possible', continued McCormick. 'We don't know what the problem is. We've got a hole in the side of the airplane and we've got a

The collapsed floor at the rear of the cabin. *(Captain Bryce McCormick)*

full left rudder here. But we're under control and we're heading, ah, two three zero at the present time, letting down slowly to nine thousand.'

'Roger, two three zero, and continue around the, ah, turn to a heading of two seven five.'

'Continue around to two seven five, sir.'

'Can you make a standard rate of descent?' asked Cleveland.

'Ah, negative, we gotta go a little slower.'

'Understand, slow descent. And what about turns?'

'Ah, turn we can give you, ah, is close to fifteen degrees maximum.'

'OK, that'll be fine.'

'I have no rudder control whatsoever, so our turns are gonna have to be very slow and cautious.'

'Understand.'

The crew discussed their predicament further and agreed that with judicious use of engine power the captain could more readily control the descent and turns. It was an amazing and fortunate coincidence that this incident had happened to a pilot who had previously practised such procedures. McCormick was thankful for his earlier experience in the simulator and was glad that he could now put the lessons learned then to good use.

'Thank goodness it's one and three we've got', said Whitney to his captain.

'American 96,' continued Cleveland, 'descend to five thousand. Say altitude now.'

'Twelve to five', replied Whitney.

'American 96, if you are able to, squawk code zero two zero zero on your transponder.'

'OK.'

'American 96, altitude now?'

'Ah, eleven thousand two hundred.'

'OK. About two hundred feet per minute, then?'

'Yes, sir.'

'American 96, you wanna make a left turn now to a heading of, ah, two four zero.'

'Two four zero on the heading?' questioned Whitney.

'Yes sir. Two four zero, so it can give you, ah, a little more room for descent.'

'Yeah. Give me plenty room to start a long approach.'

As McCormick gingerly descended the stricken aircraft he was able to experiment with the use of engine power to aid control. The attempt seemed to be effective. Number two engine appeared to be inactive but they were uncertain if it had stopped at the time of the decompression or if it was still operating in idle. Whitney and Burke executed the appropriate drill and shut the engine down. What effect it had on the

Detached ceiling panels expose the roof wiring. *(Captain Bryce McCormick)*

engine was unknown but it seemed to ease the heaviness of the elevators. In the cabin the flight attendants were equally busy with procedures as they prepared the passengers for an emergency landing. Chief Flight Attendant Smith gathered the stewardesses together on her return from reporting to the captain and gave them details of the situation. She then spoke on the PA while the other flight attendants gave a demonstration of the brace position. She instructed the passengers to lean forward when told with their seat belts tightly fastened and to place their heads on cushions on their laps with their arms folded across their heads for protection. She then pointed out the six exit locations they would use, disregarding the two rearmost doors near the damaged area, and instructed them in the use of the emergency escape slides. The passengers were then requested to remove spectacles, pens, combs and other sharp objects which could cause injury and the attendants collected them in plastic bags. Shoes were also taken off to avoid ankle and foot injuries.

With the cabin secure and the briefings delivered, Smith returned to the flight deck to inform the captain that her preparations were completed. She also asked McCormick if he would give the 'brace' command on the PA if required.

The return to Detroit was taking a little longer than anticipated and, with just over ten minutes to touchdown, the approach procedures were commenced.

'OK, give me about fifteen degrees on the flaps now', called McCormick to his co-pilot. 'Watch it carefully.'

'We'll be landing about two-ninety-two thousand pounds', added Burke.

It was possible to lower the weight further by dumping fuel overboard but since there was little in the tanks and the aft damage was unknown the thought was discarded. The apprehension on the flight deck could be felt and the air was very tense. At that moment one of the stewardesses popped her head through the cockpit door.

'Do you guys have a problem up here?' she asked.

In the strained atmosphere, with the captain struggling to maintain control, the question seemed quite ridiculous and the crew laughed.

'Yes, we have a problem', they called.

She asked if the escape slides would be used but the captain said he didn't know and that if necessary he would activate the evacuation signal.

Captain McCormick spoke once more to the passengers on the PA as calmly as possible, assuring them that the aircraft was under control and that they were returning to Detroit. He apologised for the inconvenience and said that American Airlines would do all they could to provide transport to their destinations. The composed and routine tone of his voice had a comforting effect and helped ease the tension in the cabin.

'American 96, Cleveland, call me out of ten thousand.'

'We're out of eight seven for five thousand, right?' replied Whitney.

'American 96, roger. Turn back right now, heading'll be two eight zero.'

'Two eighty. I'm guessing that we're gonna have to have a long turn on the final for the ILS. I have no control on rudder and steering directions.'

'Roger, I'm planning about twenty miles. It'll be enough?'

'Ah, I hope so, thank you. I'll keep you advised.'

'American 96, now cleared to maintain three thousand.'

'Three thousand.'

'American 96, contact Metro Approach now, one two five one five. Good night.'

'Thank you. One two five one five.'

'We've got a nice rate of descent,' commented the co-pilot to McCormick; 'even if we have to touch down this way we're doing well.'

'American 96, Detroit', called Metro Approach.

The left tailplane damage caused by the ejected cargo door. *(Captain Bryce McCormick)*

'Loud and clear, sir. And we're through fifty-five hundred for three thousand.'

'American 96, turn right heading three six zero, descend and maintain three thousand. Vector for the ILS for three left final approach course. Altimeter two nine eight five. Visibility one and one half. Braking, clear for all types of aircraft.'

'We'll probably have no brakes, ah, so we'll try reverse . . . we're heading north and we're outta forty-eight hundred for three thousand.'

'American 96, turn right heading zero two zero. Your position is thirteen miles from the marker. You're cleared for ILS three left approach.'

'Zero two zero, and cleared for ILS three left approach, American 96.'

The DC-10 joined the ILS localiser radio signal indicating the extended centre line of runway three left at about eighteen miles (twenty-nine kilometres) from touchdown, thirteen miles (twenty-one kilometres) from the marker beacon positioned at the five miles (eight kilometres) to go point. The aircraft speed was 150 knots with a rate of descent, or sink rate, of 6–700 ft (180–210 m) per minute.

'Well, gimme the gear', called McCormick.

Whitney leaned forward and selected the landing gear lever to down.

'OK,' said McCormick, 'here we're coming into the ILS. I'm gonna start slowing her down. Give me twenty-two on the flaps.'

The aircraft then intercepted the ILS glide path radio signal marking the descent profile at about ten miles (sixteen kilometres) from the threshold, and continued its approach to touchdown.

'All right, we got the green lights', commented the captain, confirming the landing gear was down and locked.

'American 96,' radioed Approach, 'you're two and a half miles from the marker, contact tower on one two one point one. Good night.'

'Good night, sir.'

Whitney switched to the tower frequency as the DC-10 neared the marker beacon.

'American 96 is approaching the outer marker inbound.'

'American 96, roger,' replied the tower, 'continue your approach.'

The DC-10 was now only three minutes from touchdown and the atmosphere on the flight deck was very tense.

'American 96, the wind is one two zero at five, cleared to land.'

'American 96,' repeated the co-pilot, 'cleared to land.'

'Give me thirty-five on the flaps', called McCormick.

With full landing flaps set and the landing gear down, the sink rate doubled to 1,500 ft (460 m) per minute, and the captain had no choice but to increase power to maintain an acceptable descent of 800 ft (240 m) per minute. The speed rose to 165 knots, thirty-five knots faster than the required threshold speed for the aircraft's 130-ton weight. The touch-down would be fast, but if they didn't land too far down the runway the DC-10 should be able to stop in the 10,500 ft (3,200 m) length available. It was necessary to fly the entire approach with the nose yawed five to ten degrees to the right to keep in line with the runway, so what would happen when the wheels touched was anyone's guess. The captain tried to reduce speed a little by pulling back on the power but the sink rate rose dramatically. Whitney was calling out the rates of descent and was alarmed to see it momentarily increase to 1,800 ft (550 m) per minute. McCormick restored the power to regain the approach path and settled for the high landing speed.

'I have no rudder to straighten it out when it hits', McCormick reminded the others.

The DC-10 crossed the runway threshold at 100 ft (thirty m) and the captain began to pull back slowly on the control column. The movement was so stiff he had to ask his co-pilot to give him a hand. McCormick squeezed on a little more power to lift the nose for a gentle touchdown. The aircraft was flying flat and fast and floated some way down the runway, but eventually the wheels contacted smoothly with the surface at 19:44, twenty-four minutes after take-off. The aircraft landed 1,900 ft (580 m) deep into the runway at a speed of 160 knots, with the nose pointing slightly to the right. McCormick was only just thinking to himself that it had been a good landing when 'all hell broke loose'.

Almost immediately Flight 96 ran towards the right side of the runway and for the first time since the incident the captain felt he had completely lost control. The DC-10 veered off the runway and ploughed through the grass. The spoilers on the wings deployed automatically to impair the lift and help the aircraft settle on the ground while the captain pulled on full reverse power on the wing engines. McCormick shouted at his co-pilot to take control of the thrust levers and grasped the flying controls with both hands, holding the wheel left in an effort to keep the wings level. Whitney pulled the left wing engine throttle to maximum reverse power and cancelled the right wing engine reverse in an attempt to guide the machine. The asymmetric reverse power had the desired effect and countered the influence of the trailing right rudder. The swing to the right stopped and the DC-10 paralleled the right side of the runway.

The captain desperately tried to brake using the toe pedals on the rudder bar but with the bar askew to the extreme left it was very difficult

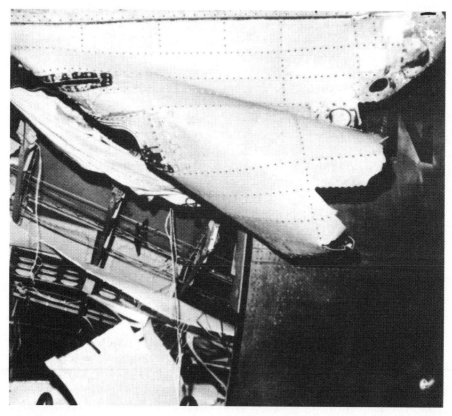

View from the left side of the fuselage showing the collapsed floor inside and the skin torn by the ejected door. *(Captain Bryce McCormick)*

to apply. The DC-10 raced on at speed between the runway and the parallel taxiway, digging up the turf in its path. Each time the aircraft met a cross taxiway McCormick tried to ease the strain on the nose wheel by pulling back on the control column, but the landing gear smashed into the hard surfaces with great force. The ground was extremely rough and with the bumpy ride the crew feared that the gear couldn't take much more punishment. If it collapsed now the flight could yet end in disaster.

Gradually the aircraft began to decelerate and the effect of the rudder displacement decreased. Under the influence of the asymmetric reverse thrust, aided by the full left deflection of the rudder bar which steered the nose wheel about ten degrees to the left, the DC-10 slowly eased to the left and back onto the paved surface. Eventually the aircraft came to a halt, severely shaken but still intact apart from the hole at the rear, about 1,700 ft (520 m) from the end of the runway. The left main landing gear and the nose wheel rested on the runway with the right gear still on the grass. McCormick, alarmed by the rough landing and fearful of ruptured fuel lines and the risk of fire, ordered an emergency evacuation. He activated the evacuation signal while the other flight crew carried out the appropriate drills. Quickly Whitney and Burke completed the check list.

'OK,' said the co-pilot, 'now the engines at your discretion.'

'SHUT 'EM DOWN!' called back the captain.

In the cabin the flight attendants rushed to their tasks and using the chutes at the forward six exits evacuated the aircraft in thirty seconds. The last of the travellers were off before the crew had finished their duties in the cockpit, and the captain could see his passengers on the left looking up at him from the far side of the runway. Around the aircraft the fire trucks were standing by. At the last moment Chief Flight Attendant Smith called into the flight deck.

'All passengers are off, captain. Goodbye. See you later.'

McCormick quickly left the cockpit and marched down the aisle of the aircraft to check that all the passengers had departed. As he passed the first door a fireman yelled at him to jump but he called back saying he still had things to do. When he reached door three left he stepped on to the wing to check for fire but could see that the aircraft was safe. He thought that number two engine would be damaged but when he looked up he could see it was intact. Then he noticed the aft left cargo hold door was missing and the damage it had caused the tailplane. He stepped back into the cabin and, joined by his co-pilot and flight engineer walking down the opposite aisle, the three made their way to the rear of the aircraft. Only now could the flight crew realise the full extent of what had happened. They could see the amount of damage the aircraft had suffered and now appreciated how well they had done to effect a safe landing. McCormick's totally professional approach to his job had won the day. They had been lucky, too, of course, but the entire crew had done a magnificent job. All passengers had escaped unhurt, except for a few minor injuries sustained during the evacuation.

With the immediate danger over the three flight crew took their time in leaving, but when they returned to the cockpit to collect their belongings they were unable to find their hats. Unknown to them, one of the stewardesses had found the hats lying on the cabin floor and had thrown them into a forward toilet. Eventually the hats were discovered and, with uniforms complete, Captain McCormick and his colleagues disembarked.

Captain McCormick and the crew received distinguished service awards from American Airlines and the captain also received a card for himself and his wife to travel first class with the airline with reserved seats. Only the president of the company can displace them. McCormick's fellow pilots also voted him president of the Grey Eagles, an organisation of senior pilots.

Chapter 5

Don't be Fuelish

Weather, followed by fuel, are the two most important factors of any flight, the forecast of the former often deciding the quantity of the latter. In the 'good old days' it was simply a matter of filling the tanks, a procedure which became somewhat modified with the introduction of jets. Even then it was standard practice to carry more than requirements, adding extra for 'mum and the kids' and a 'bit for gran' as well. Fuel was plentiful and cheap and no one thought much about it. The first oil crisis which began in the 1970s ended that attitude, with fuel prices trebling in three years. Costs to industry and the private individual soared. The United States was hit especially hard, the outcome of the crisis being smaller, more fuel-efficient cars and lower speed limits on the highways. An advertising campaign was introduced across America encouraging people to save fuel, with billboard signs saying 'Don't be fuelish' being pasted throughout the nation.

Governments of the world encouraged their public to be fuel-conscious and industry searched for any means of saving fuel. Airlines found their fuel bills soared to thirty per cent of operating costs and companies were forced to examine their fuel policies closely. The world's airlines had no alternative source of energy to which they could turn and the only option open to them was to carry less fuel. The minimum legal fuel load requirements for a journey are sufficient fuel for the flight from A to B, extra fuel for a diversion to C (in case of problems such as bad weather at B), some reserve fuel to cover contingencies, and a little for taxiing. A typical London–New York flight for a Boeing 747 would require eighty tons of fuel for the journey from Heathrow to Kennedy, thirteen tons in case of diversion, say to Boston, four tons of reserve fuel for contingencies, and one ton for taxiing. The total fuel requirement would be ninety-eight tons but a flight would normally use only eighty-one tons – trip plus taxiing fuel – and would land with seventeen tons still in the tanks.

Any fuel carried over and above the minimum fuel load is referred to as excess fuel. The B747 is a very-long-range aircraft and the maximum fuel load which can be carried in the tanks is about 140 tons, a weight equal to a fully laden Boeing 707. The equivalent volume is 39,000 gallons (177,290 litres). On a typical transatlantic journey, therefore, the average flight has forty-two tons of spare fuel capacity. The greater the aircraft weight, however, the higher the fuel consumption, and the problem for airline accountants is that any excess fuel carried above the minimum requirements has a significant portion (i.e. three per cent per hour) of that quantity used up just to carry the excess. If ten tons of excess fuel are carried on a ten-hour flight, for example, then three tons of that fuel will actually be used carrying itself. Only seven tons will be available for use over the destination. In fact, it has been calculated that to carry regularly the weight of just one small sachet of sugar increases fuel consumption by one gallon per year.

The problem for airline captains, on the other hand, is that one of the most useless things in aviation is fuel left in the refuelling bowser. Fuel on the ground may be expensive, but fuel in the air is priceless. Although, once airborne, aircraft are able to dump fuel if required – on a 747 at the rate of two tons a minute from nozzles in the wingtips – mid-air refuelling of civilian aircraft is not possible and it is essential that sufficient fuel is carried. Captains may be under pressure to carry the lowest amount of fuel commensurate with safety and the law but, with unpredictable bad weather in certain areas and frequent unexpected delays due to traffic increases in others, the dividing line between carrying sufficient fuel and running out in flight can be very thin indeed. Not surprisingly, crews still tend to err on the side of safety and, of course, if poor weather or landing delays are likely the captain has no choice but to carry excess fuel. On

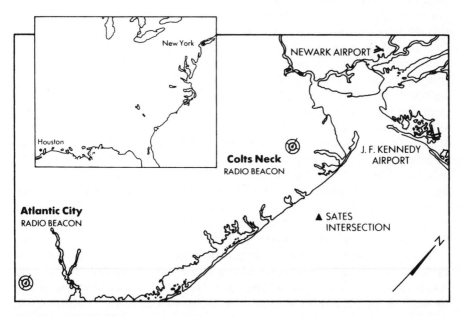

Map showing J.F. Kennedy airport, the Atlantic City and Colts Neck radio beacons, and the Sates intersection.

occasions, however, even the most conscientious of commanders can be caught out.

On 20 October 1979, a Pan Am 747 approached New York's John F. Kennedy Airport with the weather reported as poor. Low cloud and bad visibility were causing landing difficulties. The flight had departed about three hours earlier from Houston Intercontinental Airport, Texas, for the 1,200 nm journey, and now faced landing delays at its destination because of worsening conditions. The 747 had pushed back from the gate at Houston with 42.3 tons of fuel in the tanks but owing to congestion at the airport had used 1.9 tons of fuel in taxiing. The flight had eventually taken-off at the weight of 225.7 tons, with 40.4 tons of fuel remaining: 28.9 tons for the trip, 8.8 tons in case of diversion and 2.7 tons in reserve. The 747 had passed over the VOR radio beacon at Gordonsville, Virginia, cruising at 37,000 ft (11,280 m) and had been cleared to proceed direct to Waterloo VOR, situated in Delaware on Delaware Bay, 128 nm south-west of Kennedy. The captain was now further cleared to descend to 21,000 ft (6,400 m) and was instructed to join the stack at Waterloo with turns to the left, Because of the poor visibility causing landing delays at New York he would have to hold at that point before proceeding. The stacks, or holds, are flown in a precise racetrack pattern, normally over a radio beacon, with aircraft being stacked one above the other at 1,000 ft (300 m) intervals. Normally the holding procedures are conducted near airports at lower levels, so the instruction to Pan Am to hold at such a height so far out clearly indicated the backlog of flights. It looked as though it was going to be a long delay.

The captain spoke to the passengers on the public address system and told them the bad news. He had enough fuel to hold for about forty-five minutes before commencing an approach to New York, so there was no undue worry. In the meantime they would just have to go round and round until their turn came. On the descent to 21,000 ft, the crew established contact with Pan Am ground personnel on the company radio frequency and a series of messages passed between the two in an attempt to establish a landing time. It was also agreed that if Kennedy closed they would proceed to Newark, New Jersey, where the weather was known to be better. The 747 joined the holding pattern over the Waterloo beacon, still descending, and after one or two left circuits the captain was instructed to change to right-hand turns. The Pan Am aircraft changed its turns to the right and levelled at 21,000 ft.

While established in the hold awaiting onward clearance, the crew began to work out the exact fuel required for a diversion to Newark. There was every intention of landing at New York, of course, but it was as well to be prepared. At the beginning of the hold the tanks still contained fifteen tons of fuel, and using a 'flight conduct summary' table for fuel management, a diversion fuel figure of 6.5 tons was computed.

The calculations were based on an estimated landing weight at Newark of 195 tons, and using a stated distance of 25 nm between the two airports. This distance, however, was as the crow flies and, since aircraft don't normally fly in straight lines, the use of the summary tables resulted in a somewhat pared diversion fuel figure. It would be unusual, of course, for a flight to be right down to this fuel figure before commencing a diversion, so there would normally be something in hand.

After holding for fourteen minutes, the Pan Am aircraft was cleared to Sates intersection, thirty nm south of New York, via a 'Kennedy Two' arrival route, which proceeded over the radio beacon at Atlantic City. There would be a further delay at Sates after which the flight would be radar vectored to runway 22L for landing. At Sates the 747 joined the other holding aircraft. As flights were cleared from the bottom of the stack to commence their approach, Pan Am was instructed to descend lower in 1,000 ft steps. Eventually, after thirty-three minutes of holding at Sates, the 747 departed the stack on a radar heading, level at 7,000 ft (2,130 m). There were 8.2 tons of fuel remaining in the tanks and, with aircraft ahead landing safely, the captain was satisfied with their progress. The crew kept a close ear on the weather information, which indicated the visibility along the runway, or runway visual range (RVR), to be varying from 550–850 m (1,800–2,790 ft) at the landing end, so as low as 200 m (650 ft) at the rollout end. With both autopilots engaged the 747 could land automatically with an RVR as low as 400 m (1,300 ft), so there was some room to spare.

In case a go-around was required with the low fuel state, however, the flight engineer suggested preparing the aircraft for the 'minimum fuel go-around' procedures. There was a danger, if a missed approach procedure was initiated, of the fuel pumps being uncovered in the tanks by the low fuel levels. An engine could be starved of fuel and could run down. All boost pump switches were turned on and all cross-feed valves were opened. It was also recommended on the go-around not to accelerate rapidly and to reduce the nose-up angle to eight degrees.

The Pan Am flight was instructed to call Approach Control, and the first officer established contact.

'Descend to 3,000 feet and reduce to 180 knots', radioed Kennedy Approach.

The captain commenced descent and called for 5° flap to be set as he slowed the aircraft. Shortly they would be turned onto the final approach course, but before the controller could issue a heading the weather took a sudden sharp turn for the worse. The landing runway visual range dropped to 600 m (1,970 ft) and then to 300 m (985 ft). The visibility was now below the aircraft's landing limit and there was no way they could land. Nor could anyone else, for that matter, and they could hear on the radio the aircraft ahead requesting a diversion to Newark.

'Remain in line,' called back Approach, 'and continue to fly the runway course for traffic sequencing.'

Kennedy Approach then advised all aircraft that flights on the approach to land would have to be vectored along the 22L runway centre line for spacing before proceeding. The Pan Am 747 flew down the runway direction at 3,000 ft (910 m) and, with the weather still below limits, the first officer informed the controller that they would also like to go to Newark. Only 6.7 tons of fuel remained according to the total fuel weight indicator, or fuel totaliser gauge, with a calculated minimum fuel figure requirement of 6.5 tons. It was going to be close.

'Turn onto a heading of 240°,' called back Approach, 'vectors for Colts Neck, and contact Departure.'

Colts Neck VOR radio beacon lay to the south-west, about twenty-five miles (forty kilometres) in front of them, and was situated at about the same distance to the south of Newark. They appeared to be going the long way round. The first officer changed frequencies and called the next controller.

'Proceed direct to Colts Neck,' replied Departure Control, 'maintain 3,000 feet and increase to 250 knots. Leave Colts Neck heading 335° for radar vectors to runway 22L at Newark.'

Over the Colts Neck radio beacon the captain turned the aircraft to the right onto the assigned heading, but as he did so he was instructed to roll out on 320°. The aircraft proceeded on the north-westerly heading maintaining 3,000 feet and 250 knots. The 747 still had to route all the way round to the north of the airport before turning back onto a south-westerly direction for landing. The total distance covered would be nearer eighty-five than twenty-five nm. Passing several miles to the west of Newark the flight engineer scanned the indicators once more. The totaliser gauge now showed 2.7 tons of fuel remaining, about fifteen to twenty minutes' flying time to dry tanks, which would be just enough to get there with a fraction to spare. The fuel quantity gauges confirmed the amount remaining in the tanks but the fuel left seemed uncomfortably low. Again he checked the calculations. The figures tallied but the amount, even if they had gone on a direct route, hardly seemed enough. What the flight engineer did not know at the time, nor the captain or anyone else for that matter, was that the 'flight conduct summary' table on board the 747 which had been used for the fuel management computations was in error. The correct calculated fuel figure for diversion was 7.9 tons, 1.4 tons more than assumed. The Pan Am aircraft was seriously short of fuel.

Suddenly, to the flight engineer's horror, he saw the low pressure lights of the forward and aft boost pumps illuminate for number one tank. The fuel was so low the pumps were being starved of supply. The number one tank fuel gauge indicated 800 lb (360 kg), just over one third of a ton, but not much more fuel could be pumped from the bottom. The other tanks held only a little more, and he surmised that their pumps would also run dry when the fuel dropped to that level. By a quick reckoning he estimated that only 1.3 tons of the 2.7 tons indicated remaining was usable. The other 1.4 tons couldn't be taken into account because it had to be assumed it wasn't there. The pilots, already under pressure to land quickly, were shocked by the news. There was little more than five minutes' flying time remaining and a potential disaster was only moments away. The captain thought of turning quickly to the right and landing in an easterly direction on runway 11, but there were no instrument approach charts available for that runway and the aircraft was still in cloud. By now the first officer was in contact with Newark Approach and he requested their time to touchdown.

'Five minutes', replied the controller.

There was barely enough time and the captain declared an emergency. The strain on the crew was now enormous and they were under tremendous pressure. Newark Approach radar vectored the 747 directly onto the 22L runway course and within minutes they were established on the instrument landing system. Just before the outer marker the aircraft broke cloud and the runway could be seen ahead. About two minutes remained to touchdown. Newark Tower was contacted and the controller requested the nature of the emergency.

'Fuel', replied the co-pilot briefly, his heart in his mouth.

The captain, maintained the flap set at 20° and waited till he was certain of getting in before selecting the full 30° landing flap setting. After passing the outer marker the landing gear was lowered but the captain flew the aircraft high on the approach path and kept the speed about thirty knots fast. If the engines suddenly stopped through fuel starvation he hoped to be able to glide to a landing.

'What fuel do you have remaining?' asked the tower controller. With virtually nothing left there wasn't much that could be said.

'The gauges are unreliable so we don't know,' replied the first officer; 'one tank is feeding all engines.'

Approaching 200 ft (sixty m) the flight engineer informed the captain that all the fuel low pressure lights on all the boost pumps were illuminated. The 747 was moments from running out of fuel. Passing 200 ft the captain felt that a landing was certain and he called for the selection of full flap. The engines were now running on thin air. If they cut out at this moment he felt he could reach the threshold, but it would be a close-run event. The three cockpit crew held their breath in the tense atmosphere. Fifteen seconds later the wheels gently touched the runway surface and a mightily relieved crew felt the aircraft settle on terra firma. All four thrust levers were pulled into reverse and the engines responded with a roar, but it was a noise that did not last long. The aircraft slowed and the captain cancelled reverse thrust, but as he did so engine numbers one and four ran down from fuel starvation. The 747 taxied from the runway and proceeded along a parallel taxiway where, after about a mile, number two engine also ran down from lack of fuel. The captain shut down number three engine and waited for a tow truck to pull them to the terminal. In the meantime, the flight engineer checked his fuel panel and noted that the fuel totaliser indicated 1.4 tons remaining. According to the fuel quantity gauges about 1.5 tons of fuel remained in the tanks so the figures cross-checked. The four main fuel tank gauges read, from left to right, 800 lb (360 kg), 400 lb (180 kg), 1,100 lb (500 kg) and 1,100 lb again. The small amount of fuel which was indicated remaining should have been usable but was obviously not available. This discrepancy in the gauges and the error in the fuel management data of the flight conduct summary chart had placed the 747 in great jeopardy. If the engines had stopped only a few minutes earlier while the aircraft was still in cloud the 747 would have landed short of the runway and a disaster would have been inevitable. By good fortune and a quick-thinking crew a catastrophe was only narrowly averted.

The instances of airliners running out of fuel in flight are extremely rare, which is just as well, for a big jet trying to land without engine power, especially with flaps and landing gear set, has the flying properties of something close to a brick. The light aircraft pilot may be used to practising forced landings from height, usually simulating an engine failure, but such a prospect for the airline pilot is a different matter. If the incident occurs at night or in cloud the chances of success are about nil.

A big jet can, in fact, glide reasonably well and, indeed, when descent is initiated from cruise level the thrust levers are closed to flight idle and the aircraft literally becomes a giant glider. The descent rate is normally about 2,500 ft (760 m) per minute for a 747, so the rate of descent without idling engine power would be something greater. If all engines suddenly failed on a 747 at cruise height, therefore, the flight could probably remain airborne in the glide for about twenty minutes or so – as seen in Chapter 9.

At the lower levels, however, it would be necessary to select the landing gear down for an emergency powerless landing, and that would be another story. The big jet would drop like a stone. A touchdown without power is referred to as a deadstick landing and such a procedure on a big jet aircraft can be considered extremely hazardous. Great skill would be required to judge height and rate of descent and to execute a safe deadstick landing. There would only be one chance and more than a little luck would be needed as well. It would be hard to find quickly somewhere suitable to land for, unlike the light aircraft pilot, a grass field would not suffice. If a big jet suddenly lost all power in flight, attempting a forced landing and accomplishing the task safely would be very difficult indeed. Not, perhaps, entirely impossible.

On a warm afternoon of 23 July 1983, Captain Robert Pearson and his co-pilot, Maurice Quintal, boarded Air Canada's Boeing 767 wide-body jet in Montreal, Quebec. The aircraft, registered *C-GUAN*, had only just arrived from Edmonton and Ottawa and, with a fresh crew, was about to turn round and fly the same journey back.

The Boeing 767 is an ultra-modern, two-crew, twin-engined machine with advanced electronics, and the aircraft was brand new to Air Canada. The flight deck has an uncluttered appearance, with most details being presented on cathode ray tube type displays. The 767 is a pleasure to fly and is liked by pilots. On arrival in the cockpit on the afternoon of 23 July, however, Captain Pearson found they had a problem. The fuel indicating system on *C-GUAN* was faulty and there was no display of fuel load on the flight deck. The 767 fuel gauges are situated in the middle of the pilot's overhead panel and display fuel quantities in the two main wing tanks and the central auxiliary tank, as well as total fuel load. The minimum equipment list carried on board each aircraft stated that departure with this defect was not permitted. Only one main wing tank fuel gauge was allowed to be inoperative, and then the refueller had to dip the tank with a stick to confirm the level. The procedure, of course, was designed to safeguard against insufficient fuel being loaded.

In spite of the total fuel gauge failure, however, the 767 had received special dispensation to operate in that condition from Maintenance Control. Aircraft are often dispatched with systems unserviceable which are considered more of an inconvenience than a safety factor, but the minimum equipment list was designed to prevent departure in an unsafe

An Air Canada Boeing 767. *(Air Canada)*

condition. There are occasions, however, when dispensations are granted as long as proper precautions are taken.

The problem with the fuel-quantity-indicating system on Air Canada's 767, *C-GUAN*, had begun at the start of the month when a fault on the equipment was found to be intermittent. In fact, inconsistencies had been discovered months earlier on 767s operated by other airlines and Boeing had issued instructions for the systems to be checked regularly. A United Airlines 767 on a delivery flight had shown that discrepancies could exist between the quantity of fuel displayed on the gauges and the actual amount in the tanks. A situation could arise where the pilots thought they had more fuel than was actually available.

On 5 July, in Edmonton, an Air Canada certified aircraft technician, Conrad Yaremko, conducted a routine check on *C-GUAN*'s fuel gauge system. The aircraft had arrived from Toronto with a fuel processor channel unserviceable and he found that during the check all fuel quantity gauges went blank. Two fuel quantity processors on the aircraft compute fuel quantities, which are then displayed on the gauges. Each processor computer channel sums the quantities independently and the two systems then cross-check each other. If one fails the other can operate on its own.

Yaremko discovered, however, that there was a fault in the number two processor channel and that this was affecting the entire system and causing all fuel indicators to go blank. He found that if he isolated the number two processor channel computer by pulling its electrical circuit breaker he could get the system to work. All indicators would display as normal, but since there was only one processor channel operating there would be no computer cross-check of the calculated amount. It was necessary, therefore, as in the case of a main wing tank gauge failure, for a mechanic to dipstick the tanks to confirm the displayed quantity. The aircraft duly returned to Toronto and the number two processor channel was checked. Surprisingly, it was shown to be within tolerance. The circuit breaker was reset and the processor appeared to be working normally.

On 14 July, en route from Toronto to San Francisco, the number two fuel processor channel computer failed again and all fuel gauges went blank. In San Francisco the equipment in the electronics bay was re-racked and the system operated normally. On 22 July, again in Edmonton, in the evening before the day in question, *C-GUAN* arrived on another flight from Toronto. The system was operating normally but, during the routine check, the fuel gauges blanked once more. Yaremko checked the equipment and found the same number two processor channel failed again. He was unable to clear the fault but no spare fuel processor unit was available. He ordered one from the stores to be available the following evening when the flight returned once more to

Edmonton. Having encountered the trouble before he simply isolated the number two processor channel, and the fuel indications returned. In this condition the flight could depart in the morning.

Early the next day, 23 July, the 767 was satisfactorily dispatched but, before the flight left, Yaremko spoke to the departing commander, Captain John Weir. The two men agreed that the condition satisfied the minimum equipment list but that a fuel tank dipstick check would also be required. During the conversation Yaremko mentioned that on 5 July the aircraft had arrived from Toronto with the same problem and he had cleared the fault in a similar manner. In a misunderstanding, however, Weir thought he referred to the previous evening when the same aircraft had also arrived from Toronto, and the captain was left with the impression that the fault had been running from the day before.

The 767 left Edmonton on time and the flight operated via Ottawa to Montreal, arriving in the afternoon. On going off duty in Montreal, Captain Weir met Captain Pearson and First Officer Quintal on their way to work and passed on the news of the problem. He mentioned that it would be necessary to dip the tanks and added, incorrectly, that the fault had been like that from the previous afternoon. He suggested that they fuelled in Montreal for both return sectors because by doing so they would save themselves some bother during the transit in Ottawa. The conversation touched on previous problems with the fuel-indicating system in general and Captain Pearson, in a further misunderstanding, mistakenly believed that the fuel-indicating equipment was completely unserviceable. What's more he believed he was being told that it had been like that for the last day and a half.

When Captain Pearson and his co-pilot entered *C-GUAN*'s flight deck, they saw before them what they expected to see: the fuel quantity indicators were blank. But the flight had arrived with the system working and only the number two fuel processor channel computer unserviceable. This was the failure that Captain Weir had been referring to in his brief conversation with Pearson and that, of course, was why the tanks had to be dipped. In Edmonton the previous evening, Yaremko had tripped the number two processor channel electrical circuit breaker and after doing so had attached to it the relevant inoperative label. Why had the fuel indicators now gone blank? Unknown to Captain Pearson, an unfortunate sequence of events was beginning to unfold.

Before the new crew arrived at the 767, a Mr Ouellet had been sent to the aircraft by his foreman to dipstick the tanks after fuelling was completed. While waiting to perform his task he sat on the flight deck and noticed the inoperative label on the number two fuel processor channel circuit breaker. Although Ouellet was not qualified to check the system he thought he might be able to help and he reset the breaker. Immediately the fuel indicators went blank. On checking the equipment he found it deficient but, like Yaremko, he was unable to obtain a spare. He was told that one was being positioned to Edmonton for the flight's return. At about this time he was called to dipstick the tanks, and unfortunately he forgot to re-trip the number two fuel processor channel circuit breaker. The breaker remained pushed in and its position was masked by the inoperative tag which was still attached. To the eye it appeared that the breaker was tripped and all the screens remained blank.

Captain Pearson checked the minimum equipment list (MEL) which confirmed that he was not permitted to depart with the fuel-indicating system inoperative. At that time, however, the MEL on the newly introduced 767 was incomplete with many items blank and alterations taking place to it constantly. In the few months since the 767's introduction, over fifty changes had been made. The list was not considered reliable and it was Maintenance Control's practice then to authorise flights contrary to the MEL. Pearson checked the paperwork, and the fuel flight plan simply stated: FUEL QTY PROC # 2 INOP. DIP REQD. The maintenance log was also inspected and two entries were apparent. One by Yaremko in

Captain Bob Pearson.

Edmonton read: SERVICE CHK – FOUND FUEL QTY IND BLANK – FUEL QTY # 2 C/B PULLED & TAGGED – FUEL DIP REQD PRIOR TO DEP. SEE MEL. Yaremko, of course, referred to the indicators going blank during the check but that was not clear from the text. The other entry by Ouellet stated: FUEL QTY IND U/S. SUSPECT PROCESSOR UNIT AT FAULT. NIL STOCK. The maintenance log had then been signed out as satisfactory.

The captain also discussed the situation with the mechanics in Montreal and he was informed by them that a special dispensation had been received to operate in this manner. The case was supported by the chat with Captain Weir, the blank fuel gauges, the maintenance log and the belief that the machine had flown in this fashion from the day before. If others had flown it this way then he would have to do so as well, and Captain Pearson accepted the aircraft. A fuel processor was to be ready in Edmonton for the flight's return and that provided the justification for the dispensation. The idea that the blank fuel indicating system was acceptable, however, was *not* correct and the aircraft, in spite of Maintenance Control's approval, was not permitted to depart in that condition.

Captain Pearson and F/O Quintal both believed they would be operating in accordance with the applicable rules and regulations or they would not have considered proceeding with the flight. By an unfortunate sequence of events the captain had been deceived into accepting an aircraft which was unsatisfactory. There was still the fuel dip check to be performed and that would cross-check the fuel quantity on board against the volume pumped into the tanks from the bowser plus the load remaining on arrival. If the calculations were performed carefully, what could possibly go wrong? What, indeed!

Captain Pearson made the decision to carry sufficient fuel out of Montreal for the two-sector flight via Ottawa to Edmonton. The required fuel load amounted to 22.3 tons, or more correctly 22.3 tonnes – one tonne being equivalent to one metric ton or 1,000 kilograms – for the calculations were being conducted in kilograms. The system of units used in aviation throughout the world is not standardised and the present situation can only be described as messy.

The International Civil Aviation Organisation (ICAO) units are mostly metric, exceptions being distance which is stated in nautical miles and speed which is given in knots. Very few nations comply completely with the ICAO basic standard and most signatories abide by what is known as the Blue Table, with height, for example, being measured in feet. The Republic of China and Russia use metric units throughout with height expressed in metres and speed in kilometres per hour. Wind speeds are given in metres per second. Flying in a metric country with aircraft which comply with the Blue Table is not easy.

The USA uses its own modified 'imperial' system where many of the same problems apply. Almost all foreign aircraft have to convert just about everything for calculations: degrees Fahrenheit to degrees Celsius, pounds to kilograms, inches of mercury to millibars and US gallons to litres. The modern world trend, however, is towards metrication in spite of America steadfastly adhering to its own version of 'avoirdupois'. The names of units are also moving to a standardised form and are now referred to mostly by the names of people prominent in their particular field: centigrade has become Celsius, after its Swedish inventor, a cycle per second has become Hertz, after the famous German physicist; and more recently, a millibar has become a hectoPascal, after the French scientist Blaise Pascal.

In the 1970s, nations as far apart as the UK and Australia began the painful task of converting to the metric system, and in the early 1980s Canada also commenced the transition. For most countries the changeover was smooth and gradual, if not always without problems. The aviation industry welcomes any move towards a standard, if not least because the profusion of units within its own ranks was, and is, still confusing. Fuel, for example, is pumped aboard an aircraft by unit volume, and at a busy international airport like London's Heathrow, airlines ask for their fuel in litres, US gallons or imperial gallons. The fuel quantity is then converted to pounds or kilograms for trim purposes and if a dipstick check is required the tank depth is measured in centimetres or inches. The general attitude is that it is a wonder that more mistakes are not made.

Air Canada's aircraft had been refuelled for some time using litres which, in a mixture of units, were then converted into pounds. In line with the nation's transition to metric, however, the airline was in the process of changing, and the 767 fleet was the first Air Canada type to have its gauges calibrated in kilograms. On Captain Pearson's 767, *C-GUAN*, the refuellers discussing the fuel load with him were aware the amount he required was a kilogram measurement, but when the maintenance personnel calculated the quantity in litres to be pumped aboard they used the incorrect conversion factor. The conversion employed was for changing litres into pounds (1.77 instead of 0.8 for kilograms), and only sufficient fuel was loaded to give a combined litre volume in the tanks equivalent to 22,300 lb (10,115 kg). The tank quantities were checked using a dipstick which measured the depth of fuel in centimetres, which the ground personnel, using a table for conversion, then changed into litres. This volume compared with that loaded by the refueller, plus the amount in the tanks on arrival. When the captain conducted a cross-check, however, he was also given the incorrect conversion factor, and when he checked the calculations he confirmed the mathematics were correct. The refueller had also used the same figure before converting to

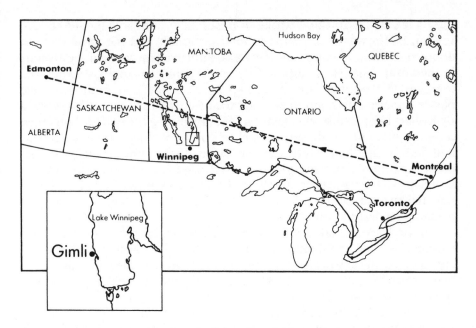

Flight 143's routeing and the position of Gimli airfield.

litres and both were in agreement. Having both used the wrong factor, however, the volume of litres in the tanks was only equivalent to 22,300 lb instead of 22,300 kilograms. One pound is less than half a kilogram, so the fuel load on board was less than half that required. With no fuel quantity indicators on the flight deck, the crew had no warning of the situation.

Before departure, the flight management computer (FMC) was loaded with the fuel quantity. A page in the Boeing operating manual at the time gave instructions which showed how fuel quantity from dipsticking the tanks could be entered in the computer. After engine start, fuel consumption would be recorded, permitting a total fuel readout to be constantly displayed throughout the flight, and en route fuel checks could be conducted normally. The FMC, however, was calibrated in kilograms and when the fuel load of 22,300 was entered, it was accepted by the computer as kilograms.

A radio call was received from the maintenance section before pushback, clearing the flight for the trip.

Captain Pearson took off from Montreal on the afternoon of Saturday 23 July operating Flight 143's first short hop to Ottawa. The aircraft arrived safely at the transit station and the turn-round was conducted expeditiously. No extra fuel was added but, as a precaution, the fuel tanks were dipped again to check the quantity before departure and to update the FMC. By an amazing stroke of misfortune, the same incorrect conversion factor was used by the ground personnel and the error was not discovered. Maintenance staff there also approved dispatch of the aircraft in the unsatisfactory condition. Flight 143 departed from Ottawa in the evening for the longer westbound sector to Edmonton. Only sixty-one passengers were on board, being looked after by six flight attendants led by Robert Desjardins, which gave a total, including the two cockpit crew, of sixty-nine. The lightly laden 767 climbed rapidly to 41,000 ft (12,500 m) and the aircraft quickly settled in the cruise. It was a fine evening, the visibility was good, and the pleasant circumstances offered a feeling of wellbeing to the passengers and crew. In the tanks, however, was less than half the fuel needed for the journey and Flight 143 had no chance of reaching its destination.

The first sector had been flown by Captain Pearson and now F/O Quintal handled the aircraft on the way to Edmonton. As Flight 143 approached the border of Ontario with Manitoba, the 767 was about halfway on its journey. At 20:00 Central Daylight Time (CDT), Quintal spoke to the passengers.

'Good evening ladies and gentlemen, this is your first officer. We're presently coming up over Red Lake and are cruising at 41,000 feet. In Edmonton it's a beautiful day.'

Below, the countryside stretched before the cockpit and Red Lake could be clearly seen.

I'm going to sit here and watch the trout swimming in the lake', joked Pearson.

Ten minutes later, as the 767 crossed the border into Manitoba, the jesting stopped. If the fuel monitoring system had been operating properly, the first sign of trouble would have been a visual and aural low fuel state warning as the fuel quantity dropped below two tons. This would have been just sufficient fuel for an immediate descent and landing. With the equipment unserviceable, however, the first indication of a problem was when the instruments indicated a low fuel pressure in the left forward pump. Beep, beep, beep, sounded the warning.

'Holy . . . !' shouted Captain Pearson.

The first officer checked out the systems.

'Something's wrong with the fuel pump.'

'Left-forward fuel pump', confirmed Pearson. 'OK, what have we got here? I hope it's just the . . . pump failing, I'll tell you that.'

Moments later a second warning light illuminated indicating low fuel pressure in another pump. The captain's first reaction was that he had a computer problem but it would be safer to have it checked before going any further. It was too much of a coincidence that two fuel pumps should fail on a brand-new aircraft, so Pearson decided to divert.

'Let's go to Winnipeg.'

Winnipeg was the nearest major airport and lay 120 nm away to the south-west. Clearance was received for the aircraft to proceed direct and to commence descent. Captain Pearson resumed control, closed the thrust levers and began the descent from 41,000 ft towards Winnipeg. Seconds later, warning lights indicating loss of fuel pressure in both right-hand pumps also illuminated. At that point the crew realised that so many pumps failing simultaneously on a brand-new aircraft was highly unlikely and that they must have a fuel problem. The chief flight attendant, Desjardins, was called to the flight deck and told to prepare the passengers for an emergency landing. A few minutes later the number one engine was starved of fuel and ran down.

'We've just lost an engine,' radioed Quintal to Winnipeg Control; 'request the fire trucks standing by for our arrival.'

Three minutes later the number two engine ran out of fuel and flamed out. The aircraft was now totally without engine power and with no fuel left in the tanks it was not possible to restart the engines. The 767 could only glide to a landing. At the moment of total engine failure the 767 was descending through 25,000 ft (7,620 m) and was sixty-five nm from Winnipeg. A small air turbine dropped from the belly of the aircraft to supply power to operate the flying controls while emergency instruments in the cockpit were fed from the batteries. The cathode ray tube screens went blank and all that was left was a standby magnetic compass, an artificial horizon, an airspeed indicator and an altimeter. The captain expected to see the compass display of the radio magnetic indicator still functioning, but it did not work. Although it was mid-evening it was still bright daylight outside and the visibility was excellent. In the cockpit, however, 'it became the blackest place in the world', Captain Pearson was to comment later. A mayday call was transmitted.

'Centre, 143, this is a mayday, and we require a vector onto the closest available runway. We are 22,000 feet on . . . with both engines failed due to, looks like fuel starvation, and we are on emergency instruments . . . now please give us a vector to the nearest runway.'

In the cabin the attendants secured galley equipment and briefed the passengers on the emergency landing and evacuation procedures. The cabin occupants were informed only that the aircraft had fuel trouble, but with both engines stopped it was obvious there was a major problem. It was very quiet. Two points were in the passengers' favour, however: Captain Pearson was a very experienced glider pilot and F/O Quintal knew the area well. He had been stationed near Winnipeg while serving in the Canadian Air Force. Their knowledge and skills would be greatly needed this day. The captain slowed the aircraft to about 220 knots which he guessed would be the best speed for a long glide. Steering the aircraft according to the controller's instructions proved to be another matter. The standby compass was situated in the centre of the flight deck at the top of the glare shield and was difficult to read. It was swinging wildly and became impossible to use for flying headings.

'So I steered by the clouds underneath us', explained the captain after the event. 'I would ask Winnipeg Centre for a heading, they would say "left twenty degrees" and I would turn left about that much, judging by the clouds, and then I'd ask Winnipeg how my heading was. Using the clouds, I kept eyeballing it.'

Judging the rate of descent and planning a descent profile proved to be the most difficult task of all. No vertical speed indicator was functioning. In an attempt to calculate a descent path, F/O Quintal repeatedly asked the controller for distances to go to Winnipeg and compared those with the altitudes indicated. At the same time he also had to complete the emergency check lists, so he was kept very busy. As the 767 descended through 14,500 ft (4,420 m) its distance to Winnipeg was forty-five nm and the aircraft was dropping much faster than either of the pilots expected. By 9,500 ft (2,900 m) the Winnipeg radar controller informed them that they still had thirty-five nm to go. Flight 143 had dropped 5,000 ft (1,520 m) in ten miles and at the most they could only glide for about another twenty nm.

'We'll never make it,' said Quintal.

A hasty replanning of the situation was required.

'What about Gimli?' suggested the co-pilot.

Gimli Air Force base was about forty nm north of Winnipeg, situated on the west coast of Lake Winnipeg, and Quintal had spent some time there during his military service. The 767 was only just near the south of the lake and they could see its waters on the right-hand side. Gimli couldn't be far away but some cloud obscured the view.

'Gimli's about twelve miles away from your position', reported the radar controller.

F/O Quintal informed his captain that the base had two parallel runways and both were long enough to use. Pearson made an immediate decision to aim for Gimli, which at that time was in their four o'clock position, and they were instructed by radar to turn right 120 degrees. Descending below the cloud, with the sun about ten degrees above the western horizon, they proceeded almost northbound under radar assistance for several miles. Suddenly Quintal spotted Gimli Airfield. The time was 20:32 CDT and in the cabin all preparations were complete. Now it was up to the captain. He had only one chance at a landing. He would also be unable to select flap without power so the landing would be fast, but before touchdown they would have to lower the landing gear. It could be dropped by the force of gravity but since it would increase their rate of descent enormously it would have to be left to the last moment. Captain Pearson then asked the controller for more details of Gimli. The Air Force base was disbanded, he was told, but the right runway was still used by light aircraft and was 6,800 ft (2,070 m) long. It would be best to go for that one.

'There will be nobody on the runway when we get there, eh . . . nothing?' questioned Pearson.

'I don't know,' replied the controller, 'I can't tell you for sure.'

Quintal could see the runway clearly ahead now and he called out to his captain.

'We're going to make Gimli OK', radioed Pearson to Winnipeg.

'Great! We show you about six miles to touchdown.'

The only problem now was that the 767 was too high and fast. Pearson could make a quick orbit but if he did so he would lose sight of the runway and he might lose too much height. There was only one thing to do and that was to side-slip. The technique is used effectively by light aircraft pilots to lose height quickly but it is not recommended on a big jet. The captain pushed on right rudder and at the same time banked the wings in the opposite direction to the left. The crossed controls had the desired effect, and as the aircraft crabbed sideways through the air the drag plunged the machine earthwards. In the cabin, the passengers felt the 767 plunge close to the ground and most people assumed the captain had lost control. On the contrary, however, the speed and height were both dropping rapidly and their situation was improving.

'Five miles to touchdown', called the controller.

'Roger,' acknowledged Quintal, 'we have the field in sight.'

With the speed back to 180 knots and sufficient height lost, the captain straightened the aircraft on the final approach. He was satisfied now that he could commit himself to a landing and he called for the landing gear to be lowered. Quintal selected the landing gear lever down but nothing happened. His heart sank. Quickly he checked the handbook but he could find no reference to free-falling the wheels. The secondary procedure for landing gear extension was, in fact, located at the back of the section on hydraulics, but by an oversight the detail had been omitted from the index. He was unable to find the relevant information so, of his own volition, he selected the alternative gear extension switch to the down position. At last he heard the gear fall down into place but when he checked the indicators he was shocked to see the nose wheel had not locked down. Frantically he searched the pages of the handbook once more in a desperate attempt to find the required drill, but before doing so the time ran out.

The captain had judged the descent beautifully and he crossed the threshold at 180 knots, about fifty knots faster than normal, at just the right height. It was then that the mistake was noticed. The right runway which had been recommended for landing was darker in texture than the other and had remained unseen as it blended with the countryside. The runway on which the 767 was about to land was the lighter, left runway.

The 767 at rest at Gimli. Apart from the damage in the nose-wheel area, no other damage was sustained. *(Captain Bob Pearson)*

Its colour had been easier to pick out from the air but, although it was longer at 7,200 ft (2,195 m), it was disused. The problem with landing on the left side was that the abandoned runway was employed over weekends as a racing car circuit and the last race of that Saturday evening had only just finished.

Captain Pearson touched down perfectly within 800 ft (245 m) of the threshold at about 175 knots but as he did so the two pilots saw to their horror that people and vehicles milled about at the far end of the runway. Children were playing and cycling in the area. Beyond the activity there were tents and caravans in which the racing drivers and their families were staying for the weekend. The 767 sped towards the gathering with no reverse power or ground spoilers available to help slow the machine. In one camper vehicle parked near the runway a racer's wife, Jo Ann Barry, was washing dishes after their evening meal. Suddenly she heard a boy shout that a jet was landing.

'I opened the camper door and there was this huge plane coming at us.'

Pearson hit the brakes hard and the aircraft reduced speed, but as it did so the unlocked nose wheel collapsed. The nose dropped to the ground and the nose wheel was forced back into the housing. Showers of sparks were thrown into the air as the nose section scraped along the ground. As it turned out, the fallen nose gear was a blessing in disguise for the friction slowed the aircraft rapidly and the 767 shuddered to a halt well short of the race meeting.

As the aircraft stopped, the forward section filled with smoke and the captain ordered an emergency evacuation. The flight attendants operated the escape slides: six in total with two at the front, two at the overwing emergency exits and two at the rear. With the nose tilted down the forward chutes sloped to the ground at a gentle angle but the rear slides were very steep. In the event, however, there was no problem with so few passengers, and all aboard evacuated quickly with only minor injuries being sustained. No flames could be seen but thick, oily smoke continued to billow from the front of the aircraft and about ten large fire extinguishers, borrowed from the Winnipeg Sports Car Club, were required to dampen the emission. The fuel tanks, when checked later, were found to be completely dry.

The incident proved a lucky escape for all concerned and, but for the skill of the pilots, the result could have been much worse. Captain Pearson had displayed his abilities to the full and had executed a brilliant forced landing.

At the summing up of the subsequent Federal Government Public Inquiry Mr Justice Lockwood said of the crew, 'Thanks to the professionalism and skill of the flight crew and of the flight attendants, the corporate and equipment deficiencies were overcome and a major disaster averted.'

Chapter 6

The Blackest Day

The day did not start too well. To begin with, Pan Am's Flight 93 was late leaving Brussels and was behind schedule on its arrival into Amsterdam. The journey had been routine but there had been no chance to make up time on the short hop. The ground staff would try their best to speed the transit in Amsterdam but the Boeing 747 was a new aircraft type for the airline and had not been in service for long. Preparing the 747 would take time. The next stop was New York, and at least 100 tons of fuel would be needed for the Atlantic crossing; just pumping the fuel quantity into the tanks would take over thirty minutes. The passenger load was light, with only 152 travelling on the Amsterdam–New York sector, so time could be saved on boarding, but the number was disappointingly low for the airline. Pan Am's brand new 747, named *Fortune*, could hold 360 people, double that of the Boeing 707, and less than half the available seating capacity would be filled. The profit margin on this flight would be thin.

The day was a cool, cloudy Sunday in northern Europe, 6 September 1970, and 49-year-old Captain John Priddy, Flight 93's commander, would be pleased to be on the way. His co-pilot was First Officer Pat Levix and his flight engineer was Julius Dzuiba, known to his colleagues as Zuby. In the cabin he had fourteen flight attendants, led by Flight Service Director John Ferruggio, bringing the total on board to 169. Among the passengers was France's new deputy delegate to the United Nations, M. Francois de la Gorce, a deadheading crew returning to New York as passengers and Captain Priddy's wife, Valerie.

At about 13:30 GMT, 15:30 local time at Amsterdam's Schiphol Airport, Flight 93 was finally ready for departure. The big jet pushed back from the gate and taxied out to the runway. The air traffic control flight clearance was given and confirmed the routeing as across the North Sea to overhead London, then westbound over the UK to begin the Atlantic crossing. Approaching the threshold Captain Priddy was instructed to hold position to await landing traffic and was told take-off clearance would be given shortly. As the crew waited for the arriving aircraft to land, a radio message was received from the control tower with the controller using Pan Am's call sign, 'Clipper'.

'Clipper 93, change to Ground Control for a minute, they have a message for you.'

The first officer re-selected the ground frequency and established contact.

'We have just been informed', radioed ground, 'that there are two passengers on your flight that were refused passage by El Al.'

'For what reason?' asked Priddy.

'We have no idea.'

It seemed just another little problem to add to those of the day for the information was too vague to be conclusive.

'If you can give us the names of the passengers we'll check them out', suggested the captain.

'The names are Diop and Gueye.'

Captain Priddy called his flight service director to the cockpit and they discussed the situation. Ferruggio did not recognise the names and he knew nothing about them. The captain felt it advisable to have a talk with the two suspects so he and his service director left the flight deck to seek them out. Meanwhile F/O Levix contacted the Pan Am operations office, known as Panops, in Amsterdam and spoke to a representative on their company frequency. Any additional information they could be given would be appreciated.

Priddy and Ferruggio descended the stairs from the upper deck to the first class cabin and the service director paged the two men on the public address (PA) system. The captain and his chief flight attendant walked through first class, right down one aisle to the back of the aircraft, then crossed over to the opposite side and walked all the way back. No one came forward. A second PA announcement was made and, this time, as Captain Priddy returned to the first class cabin, the two men revealed themselves.

'Hey, that's us', they shouted.

They were sitting in the middle of the last row of seats. Immediately it could be seen that they were young men of Middle East or Arab origin and that they were very well dressed. They seemed pleasant people, and the captain felt uncomfortable as he approached them. He didn't know what he was looking for and he felt he had no right of search. And, after all, they had paid first class fares. As tactfully as possible he explained that an alert had been received and that their names had been mentioned. The two suspects spoke very good English and they talked with the captain for a few minutes. They were polite, well mannered and helpful. Captain Priddy could see no cause for alarm.

'Well, there's been some sort of misunderstanding,' he assured them, 'but I'll either have to take you back or give you the option of being searched.'

'If you want to search us, go ahead', the men replied obligingly.

Pan Am Boeing 747. *(John Stroud)*

The captain searched their persons and belongings but nothing was found. They each had only a small Samsonite briefcase containing a few papers and it was obvious they were clean. Captain Priddy was satisfied they did not pose a threat.

'I'm sorry,' apologised the captain, 'apparently there has been some mistake.'

On returning to the flight deck, Priddy found that no further information on the suspects was available and, since the warning seemed a false alarm, he decided to depart.

Flight 93 took off from Schiphol Airport at just before 14:00 GMT (now used throughout) and climbed south-west across the North Sea, reaching level cruise approximately 25 minutes later. At about 14:30, the Pan Am aircraft flew overhead London and, with the aircraft settled comfortably in the cruise, the decision to continue seemed justified. Captain Priddy contacted Panops at Heathrow and discussed the situation with them. He explained that he had searched the two suspects but had found nothing. Suddenly the flight deck door burst open and the captain stopped mid-sentence. Standing behind the flight crew at the back of the cockpit were Diop and Gueye, each brandishing a pistol in one hand and a grenade with the pin pulled out in the other. The nearer of the two men held the first class purser before him as a shield, with an arm around her neck. Pan Am Flight 93 had been hijacked!

The day for Captain Priddy and his colleagues could now not get much worse, but just how black it had become for aviation at large was not known to Pan Am's crew. The first blow of the day was struck at 11:15 when a TWA Boeing 707 was hijacked on departure from Frankfurt, West Germany. The flight had originated in Tel Aviv and was bound for New York. The aircraft was commanded by Captain C.D. Woods who was forced by a group of armed guerrillas to fly to an unknown destination in the Middle East. Only fifteen minutes later, at 11:30, the hijacking of another Boeing 707 began, this time of airline El Al, by two passengers, a man and a woman. Their names were Patrick Arguello and Leila Khaled. Like TWA's aircraft, El Al's Flight 219 had also originated in Tel Aviv and was bound for New York, but it had, instead, transitted Amsterdam.

About thirty minutes after departure from Schiphol Airport, Arguello had let out an 'animal type bellow' and, with a small .22 pistol in his hand, he had rushed towards the front of the aircraft. He was quickly followed by Khaled, who pulled a couple of grenades from her bra and charged down the aisle with one in each hand. As Arguello approached the flight deck door he was intercepted by Steward Shlomo Vider and a scuffle ensued. Armed sky marshals raced into action but before they could do anything Vider was shot in the chest. In the first class cabin a guard lunged at Khaled but she managed to pull a pin from one of the grenades. He grabbed her elbows from behind but as he pushed her down the grenade tumbled to the floor. Miraculously it failed to explode.

Another security guard rushed forward from the rear, his pistol firing, and Arguello dropped, fatally wounded. Khaled was securely bound with string and a necktie and Flight 291's commander, Captain Uri Bar-Leb, informed London he required an emergency landing at Heathrow. The captain also radioed Tel Aviv with news of the attempted hijacking and received a request for an immediate return to Israel. The authorities there would have liked very much to get their hands on the hijacker, but Captain Bar-Leb insisted on landing in London. The wounded steward was seriously ill and urgent medical attention was needed.

On Flight 219's arrival at Heathrow a brave crew member ran from the aircraft with the unexploded grenade and placed it some distance away on the tarmac. Vider was rushed to hospital, the body of Arguello was removed and Khaled was placed in custody. The 707 was thoroughly searched and after a short delay was permitted to proceed on its way. Steward Vider later made a full recovery from his injuries.

In the afternoon, three hours after the incident, Pan Am's Flight 93 had followed El Al's route out of Amsterdam and had suffered a similar hijack attempt. The 747, however, did not carry armed sky marshals and the Pan Am big jet had been commandeered without resistance. Captain Priddy, at this stage, had no knowledge of the previous events and it was only later that the authorities were to realise how closely connected were the El Al and Pan Am hijack incidents.

As if the morning's attempts at air piracy were not enough – with one successful and one failed – another hijack occurred in the early afternoon before Pan Am became involved. At 12:15, a Swissair DC-8 en route from Zurich to New York was also overcome by a group of armed guerrillas. The hijacking had taken place near Paris and the commander, Captain Fritz Schreiber, was also forced to fly to an unknown destination in the Middle East.

Captain Priddy's 747 was the fourth aircraft that Sunday to be boarded in as many hours by armed hijackers, and 6 September 1970 was a black day indeed. It was the worst for air piracy in the history of civil aviation. Had the Pan Am captain been made more fully aware of the earlier events and had the warning of the Schiphol Airport police regarding the two suspects been more specific, he would never have left the ground. The police certainly knew of the attack on El Al before the 747 departed. The polite and helpful young gentlemen Captain Priddy had interviewed earlier now stood on his flight deck, armed and nervous guerrillas, determined to carry out their task. With the pins removed from the grenades each held, any attempt at opposition would have been suicidal and there was no choice but to obey the hijackers' demands.

Soon after the 747's takeover the guerrillas ordered Flight 93 to turn back to Amsterdam. 3100 was selected on the transponder, at that time the hijack squawk code, and London Airways requested the 747 to descend to 27,000 ft because of traffic. A few minutes later the hijackers changed their minds and decided on Beirut, the capital of the Lebanon, as their destination. Priddy received clearance to climb back to a more suitable level and continued the flight eastwards to Beirut.

'Are we going to land there?' asked Captain Priddy.

'I don't know,' said Gueye, 'I'll tell you when we get there. I have to talk to my people on the ground.'

The reply was an indication that perhaps not all had gone according to plan and further instructions were required. The 747 had plenty of fuel for the expected Atlantic crossing so the four-hour flight to Beirut in the opposite direction did not give the captain cause for concern on that account. He hoped, once on the ground in Lebanon, to be able to persuade the guerrillas to release the passengers. Unaware of the facts, he was not to realise the outcome of the events.

As the 747 flew towards Beirut the hijackers began to relax a little, but one or other of them remained on the flight deck for the entire journey. At all times the grenades were grasped with the pins removed. On one occasion one of the guerrillas played with the spare pin in his fingers and accidentally dropped it. For a while it couldn't be found and a shiver ran down F/E Dzuiba's spine at the thought of being unable to secure the grenade. The flight engineer rummaged on the floor with his flashlight and the commotion attracted Priddy's attention.

'What the hell's going on?' asked the captain.

'Never mind, keep flying,' called back Dzuiba, 'I'll find it.'

Eventually a relieved flight engineer recovered the pin and returned it to its owner.

Meanwhile in Amsterdam, the authorities had not been inactive and details of the hijackings began to be pieced together. Diop was identified as Mazn Abu Mehana, a Palestinian from Haifa, and Gueye as Samir Abdel Meguid, a Palestinian from Jerusalem. Both men, like their colleagues who had hijacked the other aircraft, were members of the Popular Front for the Liberation of Palestine (PFLP), an organisation dedicated to furthering the Palestinian cause by whatever means available. Aircraft of the enemy, Israel, or of their supporters, American or European, were considered legitimate targets.

Diop and Gueye, it transpired, had originally been booked on El Al's Flight 219 which had departed Amsterdam with Arguello and Khaled aboard a few hours ahead of the Pan Am 747. A team of four was considered the minimum required to tackle an El Al aircraft with armed sky marshals but, with the two men denied access to Flight 219 and no time to find replacements, the attempt had failed. El Al's suspicions regarding Diop and Gueye had arisen ten days earlier when they had tried to make a booking on the Amsterdam–New York flight. Security officers checked the men out and found the facts didn't tally. Both men carried Senegalese passports and claimed to be students travelling to South America. The journey they had chosen via New York to Santiago in Chile, however, was not the cheapest route to fly and, what's more, they had bought one-way first class tickets paid for in cash. Students from Senegal did not normally travel in this fashion! Diop and Gueye were eventually told that the El Al flight out of Amsterdam to New York was full and they were unable to carry them. The airline simply marked the paid-for El Al tickets with an 'open endorsement' which allowed them to fly with any other airline.

Three days before Sunday the 6th, the two men booked on Pan Am's Flight 93. At check-in, the counter clerk noted the El Al 'open endorsement' tickets but received the two men for the flight. Pan Am representatives, however, became concerned that they might have unwittingly accepted rejects from another airline and they checked the names with El Al. At about the same time, news of the attempted hijacking of Flight 219 reached Amsterdam and El Al security agents became aware of the potential danger to the 747. Pan Am's Flight 93 was already taxiing out for take-off at this stage so the Schiphol Airport police were contacted immediately. Unfortunately the police did not take the threat as seriously as the airline, and the information, when passed via the ground controller to Captain Priddy, was vague and incomplete.

When the Pan Am captain descended the stairs to interview the suspects, unknown to him at the time the two men had been sitting in the third row on the right side of first class. After the PA announcement calling for them the guerrillas had slipped back to the last row of seats in the middle. Captain Priddy had searched the right men in the wrong place.

The original intention of the PFLP was to hijack three aircraft that September Sunday: the TWA 707, the El Al 707 and the Swissair DC-8. When Diop and Gueye were found to be spare they had been switched to Pan Am to cover any possible failure of the attack on Flight 219. The plan had been carefully organised and executed and success in three out of four attempts was quite remarkable. All four aircraft had been hijacked at the beginning of transatlantic flights and carried plenty of fuel. The one problem facing Pan Am's hijackers, however, was that they expected the aircraft to be a 707. The Boeing 747 had been only newly introduced into service and it had caught them by surprise.

As Flight 93 flew towards Beirut, Diop and Gueye rearranged the seating of the passengers. With one on the flight deck at all times, the other, pistol in hand, herded all the cabin occupants into the rear of the economy section. Passports were collected and were thoroughly checked. Any potential trouble-makers – military personnel, diplomatic staff, crew members – were then isolated from the main group and seated in first class. A special watch would be kept on them. All passengers were firmly ordered to stay in their seats and to remain still. If any rescue was attempted the perpetrators would face grave consequences.

After several hours the captain was finally permitted to talk to the passengers and he informed them on the PA that the aircraft was no longer under his command. He explained, in as calm and reassuring a manner as possible, that the flight was now not going to New York, but to Beirut Airport, 2,000 miles (3,707 km) to the east of their departure point.

As the Pan Am 747 approached Beirut Airport, Captain Priddy was instructed to remain at height and to circle overhead. He obtained permission to descend to 27,000 ft (8,230 m), below the airway structure, and at the speed of 250 knots he prepared to begin orbiting. Meanwhile, in Jordan, 100 miles (160 km) to the south, another drama was unfolding. The two aircraft which had been hijacked earlier – the TWA 707 and the Swissair DC-8 – had been forced to land at a disused desert airbase in northern Jordan, known as Dawson's Field, situated near the town of Zarqa, fifteen miles (twenty-four kilometres) to the north of Amman. The landings on the rough and uneven desert strip had been hazardous and the passengers had been badly shaken. One woman aboard the TWA aircraft had sustained a broken wrist. The parked aircraft were immediately surrounded by armed commandos of the PFLP.

Detachments of Israeli airborne helicopter troops crossed the border and landed near the towns of Irbid and Jerash in preparation for an attempt to release the two airliners, but their efforts were too late. Armed factions of the PFLP also feared that the Jordanian Army, with which they had been in recent conflict, would attempt a rescue, but they too were helpless. King Hussein opposed the PFLP action in hijacking the aircraft and Jordan in fact, heavily infiltrated by Palestinian guerrillas, was on the brink of civil war. About 300 passengers, 155 aboard the DC-8 and 145 aboard the 707, were held hostage in the desert against their will.

Unaware of the circumstances, Captain Priddy began circling the 747 overhead Beirut Airport at 27,000 ft. The time was now about 18:30 GMT, 20:30 local Lebanese time, and darkness had fallen, but the lights of the town could be seen five miles (eight kilometres) below. Priddy established contact with Panops Beirut and was instructed by the hijackers to order an Arabic speaker to be brought to the mike. The guerrillas wanted to converse in their own language to conceal their discussions from the crew. An Arabic speaker was quickly acquired and he was told by one of the hijackers to summon a PFLP official to the airport to discuss the situation. A few minuted later the hijacker called back again.

'Brother, have you made contact with any of the responsible officials?'

'We are now in contact with a responsible official,' replied the Arabic speaker, 'and he shall communicate with you on this frequency when he comes to the office.'

As the arrival of the PFLP official was awaited a conflict of interest arose between Captain Priddy in the circling 747 and the Lebanese authorities at the airport. Priddy, unaware of the hostages held in the desert, was convinced the best course of action was to land in Beirut. He was confident that once on the ground he could persuade the hijackers to release the passengers, or at least the women and children. He had learned that Diop and Gueye were fairly knowledgeable about aircraft operations but their one area of weakness seemed to be in fuel management of the 747. He became aware that they had no idea of how long they could continue flying with the remaining fuel and he sensed he could force them into landing at Beirut by declaring a fuel shortage. On the ground, a Mr A. Bedran, the deputy manager of Beirut Airport, and a Colonel Salloum of the Lebanese Army, had other ideas. To them the 747 was trouble and they wanted nothing to do with it. They had their own plans for dissuading the aircraft from landing.

'For your information,' radioed Mr Bedran, 'the main runway at Beirut Airport is closed at the present time due to work in progress. The other runway cannot tolerate the weight of the Boeing 747. We suggest that the aircraft be directed to Damascus Airport, which airport is presently ready to receive this type of aircraft.'

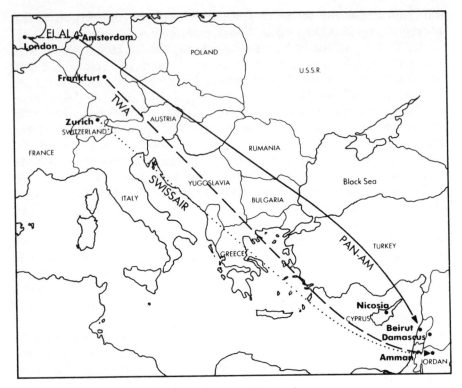

The routeings of the four aircraft hijacked on 6 September.

'We will not land at Damascus Airport, but will land at Beirut Airport. Did you hear? Answer, answer! We will land at Beirut Airport and will not land at any other airport unless contact is established with a responsible official from the Popular Front. I want an answer.'

'We are in contact – attempting to contact – the Popular Front to send a representative to the airport,' reaffirmed the deputy airport manager; 'therefore please fly overhead Beirut until contact is established and agreement reached to head to another airport as Beirut Airport is not technically ready to receive this aircraft.'

News was then received that representatives of the PFLP would arrive imminently and that they would be in the control tower within ten minutes. If the 747 crew also selected the tower frequency of 118.9 MHz on another radio box the hijackers could communicate with them shortly. It then transpired that the Palestinian officials were arriving from Bhamdoun, a mountain town some distance away, and it would take at least thirty minutes just to drive into Beirut.

An hour passed and still there was no sign of the representatives. The Pan Am 747 continued to fly in circles above the airport and by now it was obvious to all concerned that the big jet's arrival overhead Beirut was unexpected.

'It's taking a long time to get these people', commented Priddy.

'Well, you know the traffic in Beirut on a Sunday night', Gueye replied casually as if in everyday conversation.

Meanwhile Bedran, the deputy airport manager, tried again to stress that Beirut was not technically suitable to receive the 747. The main runway had been closed for over a month, the hijackers were informed, and the first 2,300 ft (700 m) of the second runway were unusable. It would be extremely difficult for the 747 to land. He did not reveal, however, that it *was* possible for the big jet to use the shortened runway for landing. Whether Bedran's remarks were bluff or fact was not known to anyone aboard the Pan Am aircraft, but when the deputy airport manager offered the information that Damascus had been contacted and the airport could receive the big jet the hijackers became hysterical. Bedran had no authority to speak to Damascus, one of the guerrillas shouted over the radio, and on no account was the 747 going anywhere until a responsible Palestinian official was contacted. Captain Priddy, however, was concerned that there might be some truth in the statements and he, in conversation with Panops Beirut on the other radio box, enquired if Baghdad would be a suitable alternative. Unfortunately, neither the airport there nor Iraqi Airways was able to handle the big jet.

On the tower frequency, the hijacker, still in conversation with Bedran, asked about one of the other events of that Sunday.

'An El Al aircraft was hijacked today en route to London. Where did it land?'

'It looks like it has landed at London', replied the deputy airport manager.

'Who is responsible for its hijacking?'

'Oh, by God, we don't know, excepting that it has landed at London at the present time.'

'Thanks.'

Diop and Gueye had, of course, more than a passing interest in the event and it appeared from what had been said that the attempt had failed. Unaware of the other successes they, too, felt that their day had not gone according to plan.

About one and a half hours after the initial request for a PFLP representative, two responsible officials, Abou-Khaled and Abou-Ahmad, finally arrived in the control tower. Abou-Khaled spoke over the radio to Gueye, now identified in Amsterdam as Samir Meguid, but neither of the men knew each other personally and both were very cautious.

'Hello, hello.'

'Yes, go ahead', said Gueye.

'Hello, hello, this is Abou-Khaled speaking, do you hear me?'

'Who?'

'Walid Kaddoura speaking, do you hear me? Hello, who is talking to me, brother? Who is talking?'

'Samir.'

'Samir?'

'Yes.'

'We have news from Amman that you could not land at the airport there', said Abou-Khaled.

'Of course we cannot land there.'

'Who is talking, brother? What squad?'

'Well . . . do you want to disclose names?'

'No, the name of the squad, the name of the squad! You originated from Amsterdam, is that right?'

'Yes, from Amsterdam.'

'What happened?' asked Abou-Khaled.

'What happened when we entered the Amsterdam Airport is that those responsible on the aircraft were suspicious and started to search us but could not find the weapons. When the aircraft took off, its destination was diverted in accordance with instructions we received.'

'Can you go to any other place? Can you go to any other place?'

'Where do you want us to take it?'

'Is there enough fuel?' questioned Abou-Khaled.

'Is there enough fuel? I will ask the captain . . . Hello, hello, he informed us that he has fuel enough for forty minutes.'

'Can you go to Amman? Is the fuel sufficient to reach the destined place?'

'Yes, it is, but landing is impossible. We can reach the destined place, but landing there is impossible.'

'Brother, if anything happens to the aircraft there, it does not matter. Is this possible?'

'It doesn't matter if anything happens to the aircraft there?' asked Gueye, surprised.

'It does not matter if anything happens to the aircraft. It did happen to the TWA's Amman landing to a certain extent, but it did not matter.'

'Does this mean that we can land there at the destined place? What I want you to understand is that this is a 747 aircraft and it weighs many times the weight of a 707 aircraft. It is impossible to land at the destined place under any circumstance whatsoever. The destined airport definitely cannot stand it, and if it lands there it will be a wreck.'

The radio conversations were in Arabic and were, of course, not understood by the Pan Am crew, but the language difference could not disguise the hijackers' concern. Not only did Captain Priddy not want to be hijacked, but the guerrillas had not wanted to commandeer a Boeing 747! If the aircraft had been a 707 it would by now be sitting at the 'destined place' – Dawson's Field – with the other two machines, but it was impossible to land such a big jet on the rough desert strip. The PFLP, therefore, were left with a rather large embarrassment circling five miles overhead Beirut: they didn't want the 747 nor did they know what to do with it.

Captain Priddy's estimate of forty minutes' fuel was on the low side and he had in fact much more than he was telling. Since Beirut Airport appeared unable to accept the 747, the hijackers suggested to the PFLP officials that they land at either Baghdad or Cairo for refuelling. With what they assumed was little fuel remaining, however, they required an answer quickly. Priddy had already established that Baghdad Airport could not handle a 747 but he did not understand the discussion in Arabic. When questioned he simply said he could remain overhead for about a further twenty minutes after which he would have to land at Beirut. If they wanted to go elsewhere they would have to decide soon.

'So I beg of you', pleaded Gueye with Abou-Khaled, 'to give an answer in less than fifteen minutes.'

'Within five minutes we will give you an answer, brother.'

By now the Beirut Airport manager had joined those in the tower and Abou-Khaled confronted him face to face. He called his bluff and demanded to know the true situation at the airport, otherwise he would order the destruction of the 747 in the air. A Colonel Hamdan of the Lebanese Army telephoned the Transport Minister, Pierre Gemayel, and relayed the threat. Could permission be given for the 747 to land? With little choice the Minister relented and allowed Beirut Airport to accept the Pan Am aircraft.

Meanwhile, in Flight 93's cockpit, the hijackers were becoming impatient.

'Hello, Abou-Ahmad, what shall we do? Have you received an answer, or not yet?'

'Abou-Ahmad talking to you. I will give you an answer in three minutes.'

'Okay, I am listening to you. Didn't you expect us here, or what? Is that why you cannot give us a quick answer? Four minutes are over, and we are still waiting.'

'They will allow you to land at the airport to refuel', a relieved Abou-Ahmad at last reported.

'Land at what airport?'

'Instructions will be given to land at Beirut Airport to refuel, which will take forty-five minutes, after which we will give you instructions as to where to go.'

'Does this mean that the Lebanese authorities agreed to our landing at Beirut? They said technically the airport was not fit for landing.'

'Technically you can land, you can land', affirmed Abou-Ahmad. 'The authorities have approved landing at the airport for refuelling.'

'Please inform them of the following message: if any attempt is made to approach the aircraft, we will destroy the aircraft completely.'

It was agreed that, on landing, the two PFLP representatives in the tower could board the aircraft for discussion, and descriptions of the men were radioed to the hijackers for identification. In the darkness it would be difficult to see who was approaching the 747. Gueye, however, was still unhappy about the fact that he was not known to those in the tower and he expressed his concern.

'I am astounded that you do not know who is talking to you as long as you know that I originated in Amsterdam.'

'Brother, it was not planned for you to land at Beirut, remonstrated Abou-Ahmad. Ashraf* and Abou-Hani* did not tell you to land at Beirut. This is an emergency. I am not supposed to know.'

'I know it is an emergency. We also did not expect it to be a 747.'

'Therefore, it *is* an emergency, and we do not know.'

Captain Priddy now prepared the aircraft for landing and the tower controller butted in to the radio conversation to pass on the relevant details.

'Okay, Clipper 93 – calling for landing – you have to be very smooth on landing and you have to choose either 03 which is landing from over the sea or runway 21 from over the sea. The full length will be available, I mean runway 03. On runway 21 the first 600 metres are to be avoided, and anyhow on runway 03 and 21 there will be a VASI** which will guide you in for landing.'

'What is the field elevation?'

'The field elevation is eighty-seven feet, sir.'

'Eighty-seven feet, okay', repeated Priddy. 'What is the wind direction and velocity?'

'Easterly at five knots, runway length available 3,180 metres.'

The Pan Am 747 had no airport charts on board for the arrival at Beirut and all relevant details had to be passed by radio, but Priddy was not over-concerned about the landing. He had already been in touch with a Pan Am unit in Frankfurt on HF long-range radio and had received much information. He was suspicious the poor conditions at Beirut had been exaggerated by the authorities to prevent a landing and he did not expect any problems.

Suddenly the PFLP officials in the tower spotted Lebanese Army vehicles and troops massing on the manoeuvring areas and they became extremely agitated. Word had also spread amongst the Palestinian community in Beirut of the excitement at the airport and a number of heavily armed PFLP commandos had positioned themselves in front

* Amman PFLP leaders
** Visual Approach Slope Indicator.

of the terminal building. An ugly scene was brewing and there was a great deal of tension.

'Do not land pending further instructions!' ordered Abou-Ahmad from the tower. 'There are army vehicles at the airport. Do not land now!'

'Delay the aircraft landing? Why?' demanded Gueye, confused.

'In case an attempt is made to approach the aircraft, they are not going to be happy at all.'

'Who is attempting to approach the aircraft? Answer! Who is attempting to approach the aircraft? Answer!'

'Can you postpone landing for the time being, brother?'

'Why? Give me convincing reasons.'

'Brother, brother. Abou-Ahmad is talking to you. We now see on the ramp army vehicles moving here and there, therefore any attempt to approach the aircraft from the army will not be tolerated by us at all. So please postpone the landing a few minutes to enable us to agree with them to pull out all army vehicles.'

'Okay, will inform the captain of this. I want to tell you something which you must accept and understand quickly, in that, if any obstacles develop and we could not land at Beirut, in about ten or fifteen minutes we will not be able to go to any other airport as the fuel would not suffice.'

'Okay, I heard you, you will definitely land at Beirut Airport for re-fuelling, but we beg you to postpone landing for a few minutes in order for us to come to an agreement with the army to withdraw all his vehicles.'

'Fifteen minutes enough?'

'Fifteen minutes in order.'

'Okay, will postpone landing for fifteen minutes.'

'Clipper 93 calling tower,' radioed Priddy.

'Go ahead.'

'Okay, this is the captain speaking. How about clearing that runway so that we can get down, huh? Do not cause us any more trouble. These folks are serious and are willing to destroy the airplane.'

'Okay, sir. We are going to give the military here orders to clear all the military around the runway so as to give way for landing safely.'

'Okay, as fast as you can, huh?'

'Okay, as fast as we can.'

'We just want to land at Beirut', restated the captain.

'Do you prefer to land on runway 03 or runway 21?'

'We'd rather land on 03.'

'Roger. Stand by, please.'

Unknown to any of the communicators, other ears were listening to the conversation. The border with Israel lies less than sixty miles (ninety-six kilometres) to the south of Beirut, and Israeli controllers were monitoring the Lebanese airport's frequencies. Suddenly a new voice was heard on the air.

'Hey, Clipper 93, this is Tel Aviv Control. Come on down! We've got two beautiful, long, lighted runways – we got emergency equipment standing by – we'll give you fighter escort. You just make a dash for it.'

The two hijackers went through the roof and Gueye became hysterical. He waved the grenade around and threatened to detonate it immediately. The Israeli controller kept trying to persuade the 747 to turn south for Tel Aviv, in spite of Captain Priddy's protestations, and Gueye was going to blow them all up on the spot. Eventually the captain managed to persuade the Israeli controller to keep quiet and he was able to pacify the hijackers. Alarmed now at how easily the situation could get out of hand Priddy stressed his concern once again.

'From Clipper 93. They say they know the airport and if there are any tricks they will blow up the airplane. So cut out the horsing around down there and get the runway cleared and get us down, huh?'

'Okay, that is what we are doing now.'

The tower controller repeatedly asked the captain for his endurance – how long he could stay in the air – but Priddy, not wanting to reveal that he still had plenty of fuel, simply kept saying that an immediate landing was required. He also radioed Panops Beirut on the other box, asking them to impress upon those concerned that the hijackers were serious in their intentions to blow up the aircraft.

'Understand', replied Panops. 'For your information, I personally saw some groups of forces, policemen and army groups, pulling out of the airport in order to avoid a clash with the men on board. I would say that it is about to be cleared.'

Ten minutes later, after almost two hours of circling, descent clearance for landing at Beirut Airport was finally received. F/O Levix changed frequency to radar control and the Pan Am aircraft was vectored on to the approach for runway 03. After a short while the 747 at last lined up on the final approach to land and a large crowd, including many armed PFLP commandos, gathered to watch its arrival.

'Clipper 93, eight miles from touchdown', informed the radar controller.

'Okay, we will stay with you for the moment . . . now we have the runway in sight.'

'Understand you have the runway in sight. Take over visually to runway 03 and call tower, frequency 118.9.'

'Okay.'

Levix changed frequency but before he could establish contact the tower controller transmitted.

'Clipper 93, Beirut tower, your range please.'

'Okay, we are about eight miles out and we have the runway in sight.'

'Roger, we have the VASI on now. Cleared to land on runway 03, wind east at five knots.'

'Okay, understand the wind is east at five knots and cleared to land, runway 03.'

'Okay, and I suggest you try to stop as soon as possible, please. We will take you to the parking.'

'Okay, I will try to run short if we can. I will stay on your frequency for taxi and instructions. Do you understand?'

'It is correct,' replied tower, 'remain on this frequency.'

'Okay.'

'East section cleared to parking. Use the first left intersection and pick up the follow-me jeep.'

'Okay, we understand that.'

'Clipper 93, this is Beirut tower.'

'Okay, go ahead.'

'After you stop, call the frequency 121.9, please, and let somebody of the people who are with you talk to us.'

'Okay.'

At 20:40 GMT, 22:40 local time, the Pan Am 747 touched down safely, if a little roughly, on the uneven surface and observers watching from the terminal clapped as it landed. A follow-me truck led the aircraft from the runway to the east parking area and as the big jet approached the terminal it turned to face away from the building. Captain Priddy set the parking brake and called for the engines to be shut down.

Abou-Khaled and Abou-Ahmad descended from the tower and mounted a jeep while another PFLP party member, Salah, maintained contact from the control tower on the ground frequency of 121.9 MHz. The hijackers could also communicate by radio with the officials in the jeep and as the vehicle approached they were ordered to stop and disembark. Anxiously the hijackers, on edge and suspicious of every move, watched from the flight deck. The jeep sped rapidly back towards the terminal and the two PFLP men approached the big jet on foot. Neither group could see clearly what was happening in the darkness but the men on the ground were accepted and the left forward door was opened. After conferring with the officials, instructions were given to the captain to order maintenance steps, which were eventually placed by the door, and the PFLP men boarded the aircraft. The door was closed and the steps were removed.

As the four men on the aircraft discussed the situation, Salah, in the tower, awaited instructions from the party members in Amman. A reply was expected in approximately half an hour regarding the fate of the Pan Am aircraft and its occupants. In due course orders were received and Salah transmitted the details to Abou-Khaled in the 747 cockpit.

'Take the following instructions', commanded Salah. 'News received from Amman to the effect that you first refuel. This means that the first thing you do is refuel, and if they refuse refuelling here, we will blow up the aircraft at Beirut Airport, of course after you disembark. Also, of course, we have notified the airport manager of this, and he is making the necessary arrangements for you to refuel. After you refuel, the same guys who are commanding the aircraft now will head on it to Cairo. In Cairo they will precisely act in like manner as Abou-Doummar and Leila Khaled have acted. Clearly, they will act the way Abou-Doummar and Leila Khaled have acted, but they have to avoid all the mistakes that occurred in the operation of Leila Khaled and Abou-Doummar at Damascus Airport. You remember what happened at Damascus Airport? – the way Abou-Doummar and Leila Khaled acted – but they have to avoid the mistakes that occurred. This means that the operation must be accomplished in perfect manner, the front and the rear, that is, the nose and the tail.'

In August of the year before, Abou-Doummar and Khaled had hijacked a TWA 707 en route from Rome to Tel Aviv via Athens. The jet had been forced to land at Damascus, where a bomb had been detonated only moments after the passengers and crew had been evacuated. Fortunately for the airline the damage was not irreparable and about one-and-a-half months later the aircraft had been returned to service. There was to be no mistake this time. The 747 was to be effectively wired with explosives at both the front and the rear of the fuselage and to be completely destroyed. The destination, Cairo, had not been chosen without reason. The Egyptian president, Nasser, had permitted peace negotiations to be conducted with the Israelis in New York and the parties were moving towards agreement. The Palestinians, however, had been excluded from the talks and the blowing up of a 747 at the Egyptian capital's airport would effectively demonstrate their displeasure.

Discussions between the PFLP men in the control tower and the 747 now centred on the refuelling of the aircraft. The hijackers were extremely nervous that the process might be used to launch a rescue attempt, and instructions were given that only the minimum number of refuellers necessary would be permitted to approach the aircraft. They were also to be closely monitored. The Lebanese Army, although withdrawn, could be quickly recalled, and the large number of armed PFLP men in the airport area created a very tense atmosphere. The situation was very delicate and in the darkness any wrong move could turn the parking apron into a battle ground. The hijackers also had a problem in explaining to Salah that they had no explosives on board and these items – codenamed 'sandwiches' – would have to be loaded. Eventually the message was understood and arrangements were made to secure the explosives from downtown.

'During this period,' radioed Abou-Khaled from the 747, 'kindly secure the items required by the guys here for travel to Cairo. There is possibility that one of us would travel with the guys to Cairo. Did you hear me, Salah?'

'I heard you, Abou-Khaled,' replied Salah, 'and have arranged everything. All the items will be available quickly. Samir El-Saheb and Captain Ali are going to supervise five other elements to refuel the aircraft. And "sandwiches" will be sent to you so that the comrades may have dinner.'

The Lebanese authorities, only wanting to see the back of the Pan Am flight, offered Captain Priddy as much fuel as he wanted. Full tanks would have been ideal, of course, but the aircraft was restricted in the fuel load it could carry because of the short distance to Cairo. The flight from Beirut to Cairo was less than a hour and the 747 would have to land at its destination at or below the maximum landing weight. A figure was agreed and refuelling began. Meanwhile, Salah passed over the radio further instructions from Amman to one of the hijackers on the 747 flight deck.

'You will execute the plan you will now work out at your end. That is to say that you, Abou-Khaled, the comrade you have, and Abou-Ahmad will work out a plan as to how you should act there. With regards to the passengers, deplane them quietly and smoothly. After the passengers deplane, determine the distribution of your responsibilities as to who goes in the front and who in the rear. The quantity of the sandwiches which we will give to you – you will amuse yourselves with them on the way – arrange them properly in order for your plan to be a complete success. This means that you will work out all the details with Abou-Ahmad and Abou-Khaled. Tell them this for me. You have received the instructions. Make out the arrangements, work out a detailed plan. That is your mission. Consider all plans previously made as cancelled. All of you aboard the aircraft work out a complete plan as to how you should act there. Secure the passengers deplaning. Of course, it is important for us that the passengers deplane safely.'

As the preparations for departure progressed the tension on the aircraft eased and the passengers were able to relax a little. Drinks were served and they were even permitted to stretch their legs on board. A political statement was read over the PA by one of the guerrillas. While waiting, the hijackers then took it in turns to walk through the cabin to talk to the passengers. Where did they come from? Had they any questions? The Palestinians' manner was friendly and polite, but the pistol in one hand and grenade in the other were readily evident. Few cabin occupants wanted to converse with them.

The explosives arrived about halfway through the refuelling process and arrangements were made to bring them to the 747. The 'sandwiches' were packed in a large seaman's bag which weighed about eighty pounds (thirty-six kilograms) and two men were needed to carry it. The maintenance steps were drawn up again to the left forward door and nine guerrillas, including one woman, brought the heavy explosives bag and a quantity of small-arms onto the aircraft. The Palestinians aboard welcomed their comrades warmly and they embraced each other by the door. When all were gathered in the 747 the door was closed and the steps were removed once more. A conference was held in first class and one of the joining group who had accompanied the refuellers to the aircraft, a Captain Ali, discussed the distribution of the 'sandwiches' on the aircraft. He was the explosives expert who was going to travel with the two hijackers on the trip to Cairo. Captain Priddy and his colleagues were waiting in first class when the group of guerrillas boarded and the meeting was held, but with all conversation in Arabic they understood nothing. Priddy asked if he could go back to talk to the passengers and he had a few words with each and every one. The captain was later followed into the cabin by several members of the guerrilla group who, like the hijackers, were friendly towards their hostages and encouraged questions. They also tried to justify their behaviour and to advertise the Palestinian cause.

By now the refuelling was nearing completion and Captain Priddy returned to the flight deck to supervise the details. He sat in the cockpit with the hijacker, Gueye, and tried to persuade him to release the passengers. Couldn't he fly to Cairo with just the crew as hostages?

'No, you don't understand', admonished Gueye.

'What will happen when we get to Cairo?'

'Everyone will get out of the airplane, I promise you', assured Gueye.

The fact that the aircraft was going to be blown up was left un-mentioned, as were a number of other details, but Gueye had never lied to the crew and the captain felt encouraged. As Priddy checked the calculations, Salah in the tower spoke some last words to the hijacker, urging him to be extremely cautious and offering him encouragement.

'Now I am going to go down,' replied Gueye, 'and I will leave the captain for a little while in the cockpit, in order to make the arrangements for the Cairo trip. This means that I will be away from the cockpit for fifteen minutes in order for him to get ready and also to send the crew to him.'

By now Diop had returned to the flight deck and he was left in charge as Gueye descended the stairs to supervise the final preparations. The two other flight crew were sent back to their posts and Captain Priddy and his team began to get the 747 ready for the flight. Departure was expected in forty-five minutes. About half an hour later, at 22:52 GMT, just before 01:00 local time on Monday morning, the refuelling was completed and the fuel bowser and ground personnel were withdrawn from the aircraft. The forward left cabin door was opened once more and a group of PFLP representatives on the ground spoke to those on the aircraft. All of the joining group were instructed to leave, except Captain Ali, the 21-year-old explosives expert, and the maintenance ladder was repositioned at the door for them to disembark.

All preparations were now completed and as the door was closed and the area around the aircraft cleared, the engines were started. A few minutes later all engines were running and at 23:30 the 747 taxied out to the runway. F/O Levix radioed for air traffic control instructions from the tower and the controller read back a flight clearance to Amman.

'We want clearance to Cairo', corrected Levix. One of the hijackers also butted in saying that they were not going to Amman.

'Well, you're cleared for take-off,' sighed the tower controller, 'for wherever you want to go.'

Approaching the runway a clearance to Cairo was finally received and one of the PFLP representatives in the tower informed the hijackers that radio communications were to cease. The 747 was to take off and proceed en route in radio silence. On no account were other stations to be informed of the aircraft's destination.

Captain Priddy started the take-off roll at 23:36 and one minute later, almost three hours after touch-down, the Pan Am aircraft lifted off for Cairo. The flight soon settled down to routine and all was as usual except for the lack of radio communications and the one hijacker who remained constantly on the flight deck. The passengers were served a meal and the children were permitted to run free. A pretence of normality returned to the situation and all on board remained calm. Only one young man, Captain Ali, busied himself on the aircraft, swiftly planting the explosives in closets, toilets and other parts of the structure. Between the charges he cut and laid the explosive fuse, a simple burning type which would have to be ignited by a naked flame to start the sequence. At the beginning he cut just sufficient for an eight-minute delay. He knew his business well and he was cool and efficient. His quiet manner masked his actions and many passengers were unaware of what he was doing. Ali concentrated on the front section of the 747 and as he worked in the first class cabin he was observed by those people segregated in that section. Amongst them was the service director, John Ferruggio, who later asked one of the hijackers what was going to happen.

'We will give you eight minutes', he answered.

'Eight minutes for what?' questioned Ferruggio.

The hijacker refused to reply.

'What about this?' asked the service director, pointing to the explosives.

'What do you care about this imperialistic piece of equipment?' sneered the guerrilla as he looked about the 747.

'I think it's a beautiful piece of equipment and, imperialistic or not, I do care.'

The hijacker paused for a moment then added matter of factly.

'It's going to be blown up. You will have eight minutes to get out.'

At this stage the 747 was passing to the south of Cyprus, in the Nicosia flight information region, and about forty-five minutes' flight time remained to Cairo. Ferruggio was concerned about the short period allowed for evacuation and he asked if he could brief the crew and passengers on procedures. He summoned all the available crew, including the four flight crew travelling passenger, to first class and told them what he knew. He suggested that the passengers be broken into groups and each section be instructed by flight attendants on evacuation procedures. The four off-duty flight crew were each asked to position by a door carrying a baby. When the chutes were deployed they were to jump first with the infant and the mother was to follow immediately. The parent was to take the child on the ground and the crew member could then help the evacuees at the bottom of the slide.

The task of dividing the passengers into groups and instructing them in evacuation procedures was carefully undertaken. An impression of urgency had to be imparted without telling everyone that from the moment the wheels touched the runway they would have only eight minutes to get out before being blown to pieces. Remarkably, much to the credit of the crew, the passengers remained calm and prepared themselves mentally for the immediate evacuation. All knew exactly what to do. They could take with them only the valuables they could wear and the ladies could carry nothing larger than a small, soft shoulder bag. All shoes were to be removed and thrown into blankets in piles for later return. When they got on the ground they were to run as fast as they could away from the aircraft.

As the 747 entered Cairo flight information region, the aircraft began descent in the darkness maintaining radio silence. The night was clear and approaching the city at 00:50 GMT, 02:50 local Egyptian time, Cairo Airport could be seen below. The tower controller tried many times to establish contact but the flight crew were forbidden to respond. By now Captain Ali had effectively completed his grim task and he had joined Gueye on the flight deck. Diop remained on guard in the cabin.

'Circle over the city', ordered Gueye.

Ali and the hijacker spoke excitedly to each other in Arabic, pointing to sights they could see on the ground. The guerrillas were taking no chances and were checking carefully that Priddy had led them to their desired destination. As they flew over the Nile in the centre of the city Gueye confirmed their position.

'OK, it's Cairo, go ahead and land.'

Captain Priddy flew the 747 onto final approach at 01:20 and Gueye broke radio silence, telling the controller in Arabic that they were about to land and confirming that the aircraft would be blown up on the ground. Cairo had already been informed from intercepted radio communications that the aircraft would probably be destroyed and they were anxious not to have their entire runway blocked. The controller asked that the 747 be stopped at the far end of the runway, and the hijacker obligingly agreed. The captain, however, had a different thought. Priddy and his colleagues had been isolated in the cockpit when the plan to blow up the aircraft had first been revealed to Ferruggio, and they had no idea of the hijacker's intentions. Captain Priddy knew, however, that the handling facilities at Cairo for a 747 would be non-existent. He felt that the best course of action would be to halt at a suitable turning point on the runway, manoeuvre without the aid of the towing truck, and taxi back to the terminal building.

On the final approach to land Captain Ali left the flight deck and returned to the forward cabin. He spoke to First Class Purser Augusta Schneider and casually asked her for a match. There was little she could do except hand him a box. Ali simply struck a light and ignited the fuse. The aircraft at this stage was still airborne and the lit fuse began to burn away time. The evacuation would have to be very quick indeed.

Captain Priddy landed safely in the darkness at 01:24, and as the aircraft rolled down the runway he began braking to stop at a suitable point.

'Go all the way to the end of the runway', instructed the tower controller. Gueye also shouted at him to go all the way but Priddy butted in on the radio.

'I want to stop short enough that I have room to turn round.'

Priddy kept braking and stopped the aircraft. Gueye screamed at him to keep moving and the controller shouted over the radio to continue to the end. In the cabin, Service Director Ferruggio felt the aircraft come to a halt and, using a megaphone, he immediately ordered the evacuation. All doors flew open and the chutes deployed satisfactorily. Instantly the dead-heading flight crew, each holding a baby, slid to the ground, followed closely by the mother and the other cabin occupants. The flight attendants were cool and efficient and the evacuation was orderly. It was just as well, for half the fuse time had already gone and only four minutes remained to detonation.

In the cockpit, Captain Priddy was still being shouted at by both Gueye and the controller to continue further down the runway and, unaware of the circumstances, he opened up the thrust levers. Service Director Ferruggio was horrified to hear the noise of the engines and to feel the 747 begin to move while people rushed down the slides. He mounted the stairs to the upper deck two at a time and from half-way up bellowed through the megaphone.

'For God's sake, stop. We're evacuating.'

Priddy jammed on the brakes and called for the emergency evacuation check list. Quickly the flight crew completed the checks, shutting down the engines and systems. Ferruggio rushed back below, the trip up and down the stairs taking only about thirty seconds, to find all the cabin occupants, including Ali and Diop, had gone. They had been evacuated in less than a minute and a half. The service director grabbed two blankets laden with shoes and threw them down the slide, then followed the bundles himself. When he hit the ground he joined the passengers running clear. By now the security forces were closing in on Ali and Diop, who were mingling with the passengers, and the two men opened fire. Ferruggio could hear the crackle of shots behind him as he ran and he feared that Gueye had shot the flight crew still sitting in the cockpit. Less than two minutes' fuse time remained.

On the flight deck, Captain Priddy and his colleagues, still very much alive, had completed their tasks and were ready for departure. Gueye still stood behind them, one eye on his watch, and he told the flight engineer, Dzuiba, to go. The two pilots rose at the same time but Gueye shouted at them.

'No, you two guys stay.'

Dzuiba left, and Priddy and Levix had no choice but to remain seated. The seconds ticked by on Gueye's watch. Were they going to be made to die with the hijacker?

'OK, go,' Gueye cried at last, 'and good luck.'

The two men rushed down the stairs leaving Gueye on his own. They quickly ran along the entire length of the 747 and back, Priddy down the right aisle and Levix down the left, to check that all the passengers had gone, but when they completed their search they found Purser Schneider still picking up shoes. Priddy shouted at her to get out and she jumped down the nearest chute while the two men slid from the left forward door. By now Gueye was in hot pursuit and he followed the two pilots to the ground. The security forces at the airport were returning the guerrillas' fire and a lot of shooting could be heard. Tracer bullets arced through the night, striking the side of the big jet. In the darkness the situation was very confusing. Priddy and Levix took to their heels and ran for their lives. Any second now the 747 would go up and they would need to be well clear to avoid injury. As the men were only twenty-two yards (twenty metres) from the fuselage, still inside the wingspan of the aircraft and fleeing past the outboard engine, the bombs planted at the front of the big jet exploded. The entire cockpit area was blown to pieces and the pilots could feel the blast in their backs, but fortunately they escaped injury. The night sky was lit up with the eruption. Moments later an enormous blast occurred at the rear, and soon after the entire top section lifted off. In seconds the 747 was a burning inferno.

Somehow everyone managed to scramble clear and only minor injuries were sustained. When the captain reached an airport bus nearby he found passengers lying on the floor, staying low to avoid the gunfire. The Egyptian driver was frozen with terror. From too close a range the

The charred remains of the brand-new 747 on the runway at Cairo the following morning. *(Captain Pat Levix)*

occupants caught glimpses of the 747 disintegrating, but the driver refused to move. Priddy quickly took over the driver's seat and drove the bus to safety. Captain Priddy's action was typical of his conduct throughout the hijack, for his behaviour had been outstanding. His calm voice over the PA and his reassuring manner in the cabin, much of the time with a pistol at his head, had sustained the passengers throughout the event and his coolness had brought them through safely. It took over an hour to round up all the passengers and crew but eventually each and every one was accounted for. The brand-new aircraft, unfortunately, was totally destroyed and only part of the tail was left standing. The three guerrillas were arrested by the security forces and seven passengers were hospitalised, while the remainder were accommodated in the airport hotel. Later that day the runway was cleared and a Pan Am 707 was dispatched to Cairo from London to collect those people fit to travel and to speed them on their way. The hijack had been a harrowing ordeal for the passengers and crew of the 747 but at least they were now free from danger. The entire crew had behaved in an exemplary fashion and their efforts had undoubtedly saved many lives.

At Dawson's Field in northern Jordan, the hostages aboard the hijacked TWA 707 and Swissair DC-8 aircraft were still being held and faced the first of many a hot and uncomfortable day in the baking desert.

That afternoon, about 100 women, children and elderly passengers were freed. In return for the safe release of the approximately 200 remaining detainees and the two airliners, the PFLP demanded the release of seven Palestinian captives: three guerrillas serving twelve-year sentences in Switzerland for a machine-gun and grenade attack in February of the previous year on an El Al aircraft at Zurich Airport, three terrorists from a rival group, the Action Organisation for the Liberation of Palestine, being detained in Bavarian jails in West Germany for a bomb attack in the same month on a bus load of El Al passengers at Munich Airport, and the lovely Leila Khaled, being held by the police in the UK.

A few days later, on 9 September, in order to press the British for the release of Khaled, a BOAC VC-10 was hijacked. The aircraft was flying en route from Bombay to London, via Bahrain and Beirut, and was commandeered as it departed Bahrain with 105 passengers and ten crew on board. The captain, Cyril Goulbourn, was forced to refuel at Beirut and was made to fly at gunpoint to the desert strip. The PFLP then held three aircraft at Dawson's Field and over 300 hostages. The airliners were surrounded by heavily armed PFLP commandos with anti-tank and anti-aircraft guns. Beyond the Palestinian positions, the Jordanian Army, helpless to interfere, ringed the defensive circle at a discreet distance with tanks and other armoured vehicles.

Representatives of the five nations involved, Britain, Israel, Switzerland, the United States and West Germany, met to discuss joint action and, with little bargaining power, moved to succumb to the PFLP demands. Hostages continued to be released in batches to join the women and children freed earlier, until by 13 September only 56 people, mostly Israeli or dual American-Israeli nationals, were still held. On that day, amid negotiations for their release and return of the aircraft, the three airliners were blown up and totally destroyed. In retaliation the Israelis arrested 450 prominent Arabs as an assurance against the safety of their nationals. Two weeks later, by 29 September, negotiations were finalised and only six hostages remained. In Amman, battles flared between Palestinian commandos and the Jordanian Army and open war erupted. The next day, President Nasser died of a heart attack. By the end of the month all the hostages had returned home safely and on 1 October the seven terrorists held by Britain, Switzerland and West Germany were freed. Hundreds of detainees and prisoners were released by the Israelis. So ended the blackest period in the history of civil aviation.

Chapter 7

Ice Cool

Alaska lies to the north-west of Canada and is America's most northern state. Its vast area stretches from 130° west at the border with British Columbia to 173° east at the far end of the Aleutian Island chain. At the Bering Strait, lying between the USSR and the USA, only fifty-six miles (ninety kilometres) of ice separate the two antagonists. Alaska, in fact, was purchased from the Soviet Union in 1867 for $7.2 m, and to this day its sale must upset the Kremlin. The land is rich in minerals, and oil is a major industry. Fishing is prominent, too, and there is also a strong military presence. The Distant Early Warning (DEW) radar line stretches from Alaska to north-eastern Canada and defends the North American continent from attack by Soviet missiles.

Alaska is a cold and mountainous land and the early pioneers who first trekked north in search of gold, and then of oil, were as rugged as the landscape they conquered. Alaska, however, is also a land of beauty, and the State boasts America's highest mountain, Mount McKinley, standing at 20,650 ft (6,294 m). The scenery, as well as the ice-cold air, takes your breath away.

Anchorage, the largest city, lies on the shores of Cook Inlet, so named after Captain Cook, who anchored in its sheltered waters on his third voyage in 1778. Hence, also, the name of the town. In the pioneering days Anchorage was a tough and wild place, but today it is a modern and sophisticated city of 200,000 inhabitants. Situated at 61°N, just south of the Arctic Circle at 66½°N, which marks the extent of perpetual daylight in the northern summer and perpetual darkness in the northern winter, the days are long in the brief summer and short in the many months of winter. The weather is harsh, being cold and snowy for most of the year, with only a three-month respite from June to August.

Travel in the rough and extensive territory is mostly by air, with many of the excellent cross-State highways being impassable for months during the winter. Light aircraft in Alaska are used as much as automobiles elsewhere. Alaska Airlines connects the north-western State with 'outside' America while many small carriers ably operate within the region. One such local airline is Reeve Aleutian Airways (RAA) which, from its base in Anchorage, serves the 10,000 inhabitants of the Aleutian Island chain. The communities are remote and stretch from the southern shore of

Alaska to Shemya Island, 1,500 nm away. RAA was started in 1932 by Bob Reeve at 'about the time that the aircraft was replacing the dog team'. Today the company is run by his son, Dick, who is owner and president of the airline. Reeve Aleutian operates nine aircraft, two Boeing 727-100s, four Electra L-188s and three Nihon YS-11s, and employes 240 staff. The 727 operations are limited to Cold Bay, which has a 10,500 ft (3,200 m) runway, constructed during World War Two, Adak, a US Navy base, and Shemya Island, a US Air Force Station at the end of the chain. The Electra and YS-11 turboprops fly the entire route network from King Salmon, near the mainland, through such places at Port Heiden and Dutch Harbor, to Shemya at the tip of the Aleutians.

Along the line of the islands, cold air from the frozen Bering Sea to the north meets warm air from the Pacific currents to the south, and intense frontal activity results. Weather conditions are difficult and changeable with fog in the summer and storms in the winter, and operations are rarely planned for night: the environment in the daytime is enough to cope with. The Aleutian Island chain, however, hooks down as far south as 53°N and the temperature variation is not as bad as might be imagined. Winter temperature lows in the islands are of the order of –7°C and summer highs are around 13°C. In the moist atmosphere, however, the temperature range is ideal for icing and it is a major hazard to aircraft.

Most freight to the Aleutians is shipped by sea, but RAA carries a lot of mail and all the airline's aircraft are in a combined cargo/passenger arrangement, with the cargo compartment at the front. At short notice the configuration can be changed, if required, to carry extra passengers.

A Reeve Aleutian Airways YS-11. *(Reeve Aleutian)*

Load factors tend to be low but, surprisingly, Reeve Aleutian carries 500,000 passengers annually, with half that total being carried in the three months of summer. The Electra is the workhorse of the airline and seems to be ideally suited to the conditions and terrain, with the YS-11 giving useful support. Both can fly in weather and to airports unacceptable to the 727. The YS-11s are costly to run and maintain, and in the combi role have only 2.2 tons cargo capacity and eighteen seats available, but the airline owns them outright and can make a useful profit from their operation.

Weather conditions, especially in winter, are a major problem, and the long experience of the airline in flying in such harsh surroundings is of paramount importance. Stormy, violent and changeable weather, bad visibility, poor landing aids, gale-force winds and icing in sub-zero temperatures all combine to form a working environment for Reeve Aleutian crews which is a tough one indeed. Fortunately the pilots are as robust as the machines they fly and as cool as the ice with which, in the long winter, they have to contend daily.

Early on the morning of 16 February 1982, a Nihon YS-11A, registered *N169RV*, was prepared to operate Reeve Aleutian Airways Flight 69 from Anchorage to Cold Bay, via King Salmon and Dutch Harbor. The aircraft had been in the hangar all the previous day and while it was there fuel had been drained from the fuel sumps to check for water. No contamination was found. At 06:00 local Alaskan time (now used throughout) on the day of departure the YS-11 was removed from the hangar and towed in the darkness to Gate 18 at the terminal. The sky was overcast at 1,200 ft (366 m) with five miles (eight kilometres) visibility in light snow, but fortunately the wind was light. The temperature indicated –21°C. The wing fuel tanks were topped up with 900 US gallons (3,400 litres) of Jet A fuel to give a total fuel load of 1,350 US gallons (5,110 litres). Once again, following standard practice, the fuel sumps were drained for a further contamination check, but no water was detected. The examination was conducted with great care for any water remaining would quickly freeze in the very cold conditions and would block filters and disrupt the fuel flow.

N169RV was a sturdy, twin-engined turboprop powered by Rolls-Royce Dart 542 engines. The aircraft was the Japanese equivalent of the HS748 and had been manufactured in 1971. At just over ten years of age it had completed only 4,385 flying hours. The machine had been acquired by RAA two years previously and had given good and reliable service. Flight 69 was carrying little cargo and the seating area had been expanded to accommodate thirty-six passengers who were travelling on the route. By 07:45 all the cabin occupants were on board and the commander, Captain Thomas Hart, prepared for departure. Captain Hart was forty-five and had 13,500 flying hours to his credit, 4,200 of

Map showing the Anchorage-King Salmon routeing.

which were on the YS-11. He was a check captain with the company and was also qualified to fly the Electra, DC-3 and C-46 aircraft. His co-pilot was 34-year-old Roger Showers. First Officer (F/O) Showers had only just converted on to the YS-11 but he had flown 6,500 hours on other types. He was a qualified flight instructor, a certified mechanic, and was also rated to fly helicopters. In the cabin one flight attendant, Cecilia Allen, looked after the passengers. She was known to the crews as Ceecee and she had kept secret that she was pregnant and was about to leave the company.

As the door was closed the co-pilot copied the clearance. The take-off runway in use was 32 and Flight 69 was cleared for an Anchorage 4 departure: after take-off the YS-11 was to climb on runway heading till 400 ft (122 m) and then to turn left onto a track of 300° magnetic. On that course the aircraft would be issued headings onto the assigned low-level airway, Victor 456, which routed via Kenai, then south-west down the Aleutian Peninsula abeam Mount Iliamna and on to King Salmon on the Naknek River. The planned cruise altitude was 12,000 ft (3,660 m).

At 07:55, in the darkness with light snow falling, Flight 69 lifted off from runway 32 at an all-up-weight of 54,220 lb (24,594 kg); 8,800 lb (3,992 kg) of fuel was contained in the wings. About ten minutes after departure the YS-11 established on airway Victor 456, climbing to

12,000 ft towards Kenai. The outside air temperature began to drop rapidly and the crew had to keep an eye on the conditions. The potential problem in the cruise was not so much with airframe and propeller icing, for the air in the frozen north in winter is mainly free of moisture, but with the freezing of water droplets suspended in the fuel.

Checks of the fuel for free visible water had, of course, been conducted in the hangar, and after refuelling, but it was impossible, using such a method, to detect in the fuel dissolved water or water suspended as submicroscopic particles. Such dissolved or suspended water is usually present in jet fuel but its extremely small quantity does not affect normal combustion. When the fuel in the tanks is subjected to very cold temperatures at cruise altitude, however, it is a potential source of ice in the fuel system. Tiny ice crystals can form which can block fuel filters and the engines can stop with the disrupted fuel flow. To prevent fuel icing, therefore, fuel heating systems are installed on most aircraft.

On the YS-11 each engine is supplied with fuel from its respective wing tank. An engine-driven fuel pump first draws fuel from the tank via a fuel heater, which is a fuel-air heat exchanger using hot compressed air bled straight from the engine compressor. The heated fuel is then fed through a filter and into a pump from where it is delivered at high pressure to the

Captain Tom Hart.

fuel control unit. The metered fuel is then directed to fuel nozzles in the combustion chambers. The fuel heater is controlled by a three-position switch on the flight deck, marked manual, off and auto. The normal setting for the switch is auto.

Any icing of the fuel filter results in a blockage which causes a difference in pressure between the inlet to the fuel heater and the outlet of the fuel filter. When the pressure differential reaches a certain limit a detector illuminates a warning light which is positioned by the fuel heat switch. With the switch selected to auto, compressor air is automatically bled from the engine and fed to the fuel heater. The fuel heat switch 'auto' position, therefore, is a safety setting designed to operate with inadvertent icing. With the switch in the manual position, compressor air is fed continuously to the fuel heater.

As the YS-11 climbed towards its cruising altitude a request was made to climb to 14,000 ft (4,270 m) and Anchorage control approved. Approaching 10,000 ft (3,050 m), however, passing over the Kenai VOR, the crew noted that the OAT was –25°C, lower than expected, and F/O Showers moved the fuel heater selector switches from auto to manual. Immediately a rise in the fuel temperatures was checked on the gauges as the fuel heaters began to operate. A few minutes later Flight 69 levelled in the cruise at 14,000 ft and, with the temperature now indicating –37°C, the fuel heaters remained on. A serious risk of fuel icing existed without their use. Some thin stratus cloud was encountered en route but no airframe icing was evident.

Flight 69's cruise towards King Salmon was routine, except for the very low temperatures, and by over half-way along the journey, abeam Mount Iliamna and the Iliamna Lakes, the OAT had dropped to –40°C. Later the temperature rose slightly to –38°C, but it was still well below normal for that altitude. F/O Showers established contact with King Salmon Approach at fifty miles (eighty kilometres) out and descent clearance was requested.

'Roger, Reeve 69,' radioed back Approach, 'descend and maintain 7,000.'

After forty-five minutes in the cruise, at 08:50 and forty nm north-east of King Salmon, Captain Hart eased back the throttles to a reduced setting. The descent from 14,000 ft was commenced at a speed of 240 knots with constant engine power.

King Salmon Airport is used by both civilian and military traffic and the responsibilities are shared; the Federal Aviation Administration provides air traffic control personnel and the US Air Force supplies the fire and rescue services. There are two paved runways, 11/29 being the main runway at 8,500 ft (2,590 m) long and 18/36 the shorter at 5,000 ft (1,525 m). The airport lies almost at sea level and is bounded to the south and west by the meandering Naknek River which in winter freezes in parts to a depth of three feet (0.9 m). The weather in King Salmon was good and on the descent the crew copied the details from the automatic terminal information service (ATIS). The wind was 320° at ten knots and runway 29 was in use. Flight 69 was heading south-west and the aircraft would simply be able to position on a right base leg before turning right through 90° onto finals to land towards the north-west. The cloud cover was scattered at 5,000 ft and the visibility was a good fifty miles. The altimeter setting was 29.32 inches of mercury.

The crew set and checked their altimeters and approaching 7,000 ft (2,134 m) the captain advanced the throttles and held the aircraft level with the speed still steady at 240 knots. After one minute, Reeve Flight 69 was instructed to descend to 5,000 ft. Again Captain Hart eased back on the power and at the same speed slowly descended the YS-11 to 5,000 ft. The captain called for the descent and approach check lists and the drills were completed. Flight Attendant Allen reported that all the passengers were strapped in and that the cabin was ready for landing. Captain Hart set the power to 12,500 rpm and 850 lb (385 kg) per hour fuel flow, and began to decrease speed to 210 knots. The controller was slow in clearing the aircraft for further descent and the YS-11 was becoming a little high on the approach. The time was now just before 09:00 and the half-light of the northern winter day had dawned, but the visibility was very clear in the unpolluted atmosphere. As Flight 69 neared 5,000 ft the aircraft was positioned ten to fifteen miles (sixteen to twenty-four kilometres) to the north-east of King Salmon and the airport could be seen clearly ahead and to the right.

'Approach,' radioed Showers, 'we're just levelling 5,000 feet and we have the airport in sight.'

'Roger, Flight 69, you're cleared for right base for a visual approach to runway 29.'

The approach controller's instruction automatically cancelled the YS-11's flight plan and Captain Hart was now free to control his own progress. He could judge his own heights and speeds and could position the aircraft as he desired for approach and landing by visually sighting the airport. The YS-11 continued descent at 210 knots and at just after 09:00, as the aircraft passed 2,600 ft (792 m), five miles (eight kilometres) from touchdown, instructions were given to contact the tower.

'King Salmon Tower, Reeve 69,' called Showers, 'five miles right base for 29, been cleared for a visual.'

'69, King Salmon Tower, roger. Report turning final, runway 29.'

'69.'

'Three zero at one zero', called back the tower, giving the wind velocity, but making a slip.

'Three zero at one zero?' questioned Captain Hart.

'Say again the last part of transmission', radioed Showers.

'Roger, the wind at this time is three two zero at one zero, you are in sight and cleared to land runway 29.'

'Three two zero at one zero,' said F/O Showers to his captain, 'that was in the ATIS.'

'Thought he was giving us a squawk', admitted Hart, referring to the four-figure code normally issued by radar controllers for setting on the transponder. 'He's not very clear today,' he continued, 'sounds like he's talking through a wet sock.'

'No, it's not very clear', confirmed Showers.

The YS-11 was now two to three minutes from landing and, with the aircraft a bit high and fast for the short distance remaining, Captain Hart levelled off at 1,800 ft (549 m) and let the speed drop. The power setting was left constant but the speed fell to below 200 knots and continued to decrease. Height and speed would also be lost on the 90° right turn onto finals, so the situation did not pose a problem. Meanwhile, the co-pilot began to perform the before-landing check. The list would not be completed till the landing gear was down and locked but some items could be initiated by the co-pilot of his own volition. Once the gear was set he would read the check list in a challenge and response manner whereby he would call the items and both pilots would confirm the tasks accomplished. The sequence of the before-landing check list was as follows and, for visual approaches, was to be performed upon entering the traffic pattern and completed before turning onto final approach, except for the final flap settings.

Item	Response
Landing Gear	Down and three greens
High pressure (HP) fuel cocks	High stop withdrawal lock (HSWL)
Landing lights	On below 165 knots
Fuel trim	Set
Prop lights	3 on and 3 off
Fuel heaters	Off
Flaps	Set
Water/methanol	On
Spill valves	Manual

Most of the items, such as the landing gear, landing lights and flaps, etc., would be set in the normal flow of events but at this stage the first officer could set the HP cock to HSWL and could switch off the fuel heaters. The 'high stop' referred to in the second item above is a mechanical propeller pitch stop which is a safety device engaged during take-off and cruise. The propellers are variable pitch and if, for example, an engine fails at take-off, the stop prevents the windmilling propeller blades from decreasing to a flat pitch, i.e. from turning flat to the airflow, and so creating enormous drag. If such an incident happened during take-off or in the cruise, control of the aircraft could be lost, and the stop is designed to prevent its occurrence. The stop is removed during the before-landing check to permit finer pitch angles for landing and is accomplished by moving the HP cock to the high stop withdrawal lock. Two blue lights then illuminate to confirm the high stops have been removed.

Another selection for the HP cock is the 'feather' position, which is used with an engine failure to reduce the windmilling propeller drag to a minimum. The HP cock is first selected to 'feather' to cut off the fuel to the engine and this is followed by the pressing of a 'feather' button. The propeller blades are then hydraulically driven in line with the fore and aft axis of the aircraft to allow the blades to cut through the air like a knife, reducing drag to a minimum.

The selection of fuel heat to off as part of the before-landing checks is also a precautionary measure. As mentioned previously, hot air is bled directly from the engine compressors to feed the fuel heaters, and extracting air in this manner causes a reduction in engine power. At the full power setting, for example, power is reduced by four per cent when compressor air is bled for fuel heat. For maximum power availability in the event of a go-around, therefore, the fuel heaters are turned off during the approach to land.

With the HP cock selected to the high stop withdrawal lock and the fuel heaters switched off, Showers routinely scanned the engine instruments. The aircraft was now two minutes from landing; the first officer had performed the drills properly and the two pilots had complied correctly with procedures. Unknown to the crew, however, the fuel in the tanks, in spite of stringent checks, had been contaminated with more than the usual minute quantities of water. The amount would probably have remained unnoticed in normal circumstances but, operating the YS-11 in such extreme temperatures, fuel freezing problems were likely, even though the water content was very small and the fuel heaters were switched off at a very late stage. During the flight, with temperatures down to –40°C, the fuel in the wing tanks would have been very cold because of the long exposure to such icy air. As long as fuel heat was supplied any ice crystals in the system were melted and easily passed through to the engine as liquid. The fuel heater on this occasion, however, only raised the temperature of the fuel entering the system to slightly above freezing. As soon as the fuel heaters were turned off, the microscopic water droplets suspended in the cold fuel began to freeze. Ice crystals began to form in the small orifices and screens in the fuel control units and pumps, and began to impregnate the fuel filters and to restrict the fuel flow.

'Barrier one,' called the tower controller to another aircraft, 'what's your position now . . . ah, I've got you in sight. Remain clear of runway one one and two nine, landing traffic a YS-11 on right base.'

After a few seconds 'tower' informed both flights that men were working near the runway.

'Barrier one and Reeve 69, there will be men and equipment on the departure end of the runway on the left side. They are clear and outside the lights.'

First Officer Showers ignored the radio call for, as the controller spoke, he could see that there was a problem with the right-hand, number two, engine. The torque pressure indicator was dropping, as was the fuel flow. Approaching 09:02, with the height just under 1,800 ft (550 m) and the speed dropping below 190 knots, Captain Hart felt the aircraft yaw as power dropped on number two engine and the nose swung to the right. He pushed on the left rudder to hold the machine straight.

'We've lost one!' shouted Showers.

The number two fuel flow was well below 500 lb (227 kg) per hour when it should have been 850 lb (385 kg) per hour.

'Fuel flow . . .', he called, pointing to the gauge.

The number two torque pressure was also down, indicating forty lb per square inch (psi) (2.8 kg/cm^2) instead of the 100 psi (7 kg/cm^2) it should have read with the throttle position at 12,500 rpm.

'Torque's very down.'

The captain advanced the number two throttle but there was no response. It was obvious the engine had failed. He turned on both relight switches and pushed the throttle forward once more, but to no avail. With the little extra height and speed in hand the YS-11 was, fortunately, in a good position for a single-engine approach and landing. The captain felt it was better to eliminate the drag from the right engine, thereby easing the control difficulties, rather than continue attempts to restart it, and he decided to shut the engine down.

'Okay, feather the . . .', called Hart.

Showers selected the number two HP cock to 'feather' and pressed the 'feather' button. Fuel was cut to the engine and as it ran to a stop the propeller blades moved to the feather position.

'Feathered?' questioned the captain.

'It's feathered.'

Neither of the fuel filter differential pressure warning lights, which would have indicated a filter blockage, illuminated, nor did the low fuel pressure warning lights. Since the crew had complied properly with procedures the two pilots did not at first associate the engine failure with icing.

Feathering of the number two engine was completed as the speed dropped below 170 knots, and under the circumstances the YS-11 seemed safe. The aircraft was still on base leg at just below 1,800 ft and as the speed dropped to 165 knots the captain continued the descent.

'Tell 'em we've got one shut down,' said Hart, 'and would like the fire trucks out.'

'And, ah, Anchorage King Salmon, Reeve 69,' radioed Showers, 'ah, get the fire trucks out. We've lost one engine on . . . we're turning final at this time.'

At this stage the flaps were selected to the approach flap setting and, with the YS-11 still a little high, Hart called for the landing gear to be lowered. As the wheels dropped he advanced the number one throttle to increase power on the left engine to stabilise the approach. Suddenly popping sounds could be heard from the number one engine and a smell of smoke could be detected on the flight deck.

'. . . it!' exclaimed the captain, 'we're losing the other one.'

Fuel icing was also affecting the left engine but in a different manner. As the throttle was moved forward, ice crystals in the fuel control servo system caused large fluctuation in the fuel flow. Blocked orifices or screens in the system were preventing the fuel pump output from matching the throttle position and too much fuel was flowing into the engine. Excessive fuel was being admitted to the combustion chambers and an intense rise in heat was occurring without a corresponding increase in the propeller rpm to drive air through the engine for cooling. The insufficient cooling airflow was causing a severe over-temperature condition in the turbine and the engine was literally burning itself out.

The number one engine power began to drop as the popping sounds continued and the smell of smoke on the flight deck increased. The engine fire was close to the air conditioning duct for the cabin and fumes were being drawn into the aircraft. No engine fire warning was given. Immediately Captain Hart began the right turn to line up on the runway. At 1,200 ft (366 m) his height and position seemed good for a safe landing but the rate of descent increased rapidly and as a precaution the captain brought the flaps up.

'Say your fuel aboard and persons', requested tower.

The controller was obviously trying to obtain the information for the emergency services but Showers was too busy to respond. As the wheels fell into place and were checked locked the co-pilot confirmed their condition.

'Gear down and three greens', he called to his captain.

The number one engine now began to surge and the power dropped and then increased at random. With the number two engine shut down and the landing gear lowered the rate of descent increased dramatically.

'Okay,' said the captain, 'tell the girl in the back we've got a problem.'

Showers pressed the call button to the intercom in the cabin.

'Hello', replied Flight Attendant Allen.

'Yeah,' said Showers, 'we've got a little problem here. We've shut down one engine and we're losing the other one. Prepare the passengers for an emergency landing and evacuation.'

Cecilia Allen went about her duties and as she did so the tower controller radioed with permission to land.

'Reeve 69, cleared to land runway two niner. The ah, emergency equipment has been advised.'

'69, roger.'

'Get the other one going', said the captain to his co-pilot.

The number one engine continued to pop and surge and the YS-11 was losing height rapidly. Large fluctuations were noticed in the left engine fuel flow, approximately centred around 1,000 lb (454 kg) per hour, and at drops in the power the captain could feel the aircraft yaw to the left. The number one engine power now began to fade and there was a danger they might not reach the runway.

'Get the other engine going,' urged the captain; 'get it going.'

The first officer had already moved the number two engine HP cock lever forward to the start position and he was in the process of pulling the 'feather' button to unfeather the engine. They were quickly running out of time but the number two engine resolutely refused to start. With both props now unfeathered and the landing gear lowered the drag on the aircraft was enormous. Rapidly the YS-11 dropped towards the ground.

'We're not going to make it,' stated the captain simply, 'we're not going to make it.'

Keeping a cool head and thinking quickly, Captain Hart turned the aircraft 90° to the left and headed south-west towards the Naknek River. There was no time now to do anything except land on the frozen surface. That, however, was going to be easier said than done, for not all the river was frozen. At the sharp bend in the river to the south of the airport the stream slowed sufficiently for the river to freeze to a depth of about three ft (0.9 m) in the extreme temperatures. About one mile (1.6 kilometres) up and down stream from this point, however, the water was still unfrozen, and only one week earlier, in warmer conditions, a water channel had been open along the length of the river. Captain Hart aimed for the two-mile (3.2 kilometres) stretch of ice.

'69 has lost both engines,' radioed F/O Showers, 'on final here to the river.'

'69, roger', replied the controller, as he watched the stricken aircraft dive to the south of the airport. Immediately he alerted the fire and rescue services and directed them toward the Naknek River.

Captain Hart aligned the YS-11 with the banks of the ice-covered river and, with the speed about 120 knots, prepared for a rough landing.

'You want the gear up?' asked the co-pilot.

If a wheel caught in a rut on the ice it could tear off the undercarriage, but the smooth belly of the aircraft would glide easily over the frozen surface. There was also another consideration: the ice in the centre of the river was clear and appeared green in colour as the water could be seen flowing below the surface. It did not look to either pilot as if the ice would hold but there was nowhere else to go. If the surface gave way it would be better to land with the gear retracted rather than have an undercarriage leg smash through the ice.

'Yeah, put it up,' said Hart.

The first officer selected the landing gear lever to the up position but the retraction was slow because the hydraulic pumps were not operating fully with the engines windmilling. At the same time the captain selected both HP cocks to 'feather'. The right prop feathered immediately but the left side was slow to respond and Hart had a great deal of difficulty in controlling the machine. Both hands were needed to fly the aircraft. As the YS-11 rapidly approached the icy surface, the ground proximity warning system sounded its alarm.

'Pull up, pull up, pull up . . .'

Captain Hart wrestled to level the wings but the was unable to raise the left wingtip. With great skill he managed to pull the powerless machine out of the dive but just before landing the left aileron struck the ice. Fortunately little damage was caused. A moment later, with the gear still retracting, Hart touched down on the frozen river as gently as possible. The belly of the aircraft gradually settled onto the surface as the wheels retracted completely and the aircraft began to slide across the ice. Both props were now in the feathered position but with the right side stopped and the left one still turning. Immediately the prop blades on both sides dug into the ice and slashed the surface. The aircraft bumped across the ice, which was remarkably smooth, and banging and scraping sounds could be heard in the cabin. The nose-gear doors were crushed, the fuselage belly skin was buckled in a number of places, but fortunately did not

The YS-11 back on its feet. Note the smooth surface of the ice. *(Reeve Aleutian)*

rip from the airframe, and the rotating safety beacon and several antennas were broken and bent. The YS-11 continued to slide and rumble over the ice. Captain Hart, still concerned that the ice might give way, applied right rudder and steered the aircraft towards a sand bank on the north side. Slowly the machine curved in an arc to the right. Suddenly, a red warning light illuminated and a bell rang indicating a fire in number one engine.

'We got a fire', called the captain.

The required drill was to place the HP cock to the 'feather' position to cut off fuel at the control unit and to pull the fuel shut-off 'T' handle to cut off fuel at the engine, as well as to isolate the services. The fuel cocks were already in the 'feather' position so it was simply a matter of pulling the 'T' handle and pressing the buttons.

'Pull 'em both?' asked Showers.

The captain nodded and the first officer pulled the fuel shut off handles.

'Gang bar, fire bottles', said Hart.

Showers pushed up the electrical gang bar and shut off the aircraft electrics.

'Fire bottles', repeated the captain as the aircraft continued to bump over the ice.

The first officer pressed both fire buttons and discharged the extinguishant from the bottles into the engines.

As the YS-11 skidded over the frozen river it continued in a gentle curve to the right. After sliding for about half a mile (800 metres) it eventually came to an abrupt stop facing towards the west with the left wingtip resting on the surface. The aircraft still weighed in excess of twenty tons and, in spite of the three feet of frozen water covering the river, the ice began to crack as it settled. Immediately, Captain Hart ordered an emergency evacuation. The number one engine was still burning in spite of their efforts and, with the possibility of ruptured fuel tanks, there was a serious risk of fire. Showers quickly left his seat while the captain remained to secure the flight deck. The co-pilot ran to the back of the cabin and assisted Flight Attendant Allen in evacuating passengers via the aft cabin door on the right side.

In the cockpit Captain Hart completed his tasks then quickly radioed King Salmon tower with news of their predicament.

'Hey, this is Reeve 69', called Hart. 'We're down on the ice, nobody's hurt, we're still on fire over here, though. And, ah, we had a fire in the air and lost power on the engines and couldn't get to the end of the runway. We had to make a quick left turn here but we're still on fire.'

Unknown to Captain Hart, help was already on the way and the fire trucks were speeding to the scene. Hart rapidly left the flight deck and went back to the cabin, where he found some passengers struggling to

The YS-11 being towed up the river bank on the specially constructed ice ramp. *(Reeve Aleutian)*

open the right over-wing exit. Flames could be seen on the left side through the passenger windows and smoke was entering the cabin. It was imperative to get the people out quickly. The captain swiftly released the catch of the over-wing exit, pulled the handle, and threw the door out of the opening. He then assisted the passengers in evacuating by that escape route. In a matter of moments all thirty-six passengers and three crew were out of the aircraft and standing safely on the ice. Only one passenger received a minor knee injury while fleeing from the aircraft. Almost immediately the fire trucks also appeared on the scene and the fire in the number one engine was quickly extinguished. Fortunately the frozen surface, in spite of the cracking, held the additional weight. In due course transport arrived to take the passengers and crew to the terminal building, but not before some people suffered frostbite in the severe temperature.

A rescue operation was mounted to save the YS-11 and stakes were driven into the frozen river to secure the aircraft to prevent its movement on the shifting ice. Air bags were then placed below the wings and nose and inflated in turn to raise the machine to a sufficient height for the lowering of the landing gear. The YS-11 was then towed three miles (4.8 km) across the ice and was hauled from the frozen surface up a specially constructed ice ramp on to the river bank. From there it was moved to the airport for repair. The entire rescue operation took five days and was only just completed in time, for the next day the ice river melted. Two weeks later, YS-11 *N169RV* flew out of King Salmon

Airport, an engineering feat almost as remarkable as the landing on the ice. The cause of the water contamination in the fuel was never discovered.

The incident had developed very quickly and a normal approach had swiftly developed into a potentially disastrous situation. Keeping cool heads, Captain Hart and First Officer Showers had reacted calmly and efficiently to the emergency and the captain's quick thinking and great skill had saved his passengers and aircraft from an almost certain catastrophe. Both pilots received Superior Airmanship Awards from the American Airline Pilots' Association.

Chapter 8

Roll Out the Barrel

Aircraft manufacture is a highly refined engineering art and modern jets are built to be extremely robust. The strength of contemporary airframes lies in their monocoque structure, whereby the skin takes much of the load although supported by a reinforcing frame. The central section of the aircraft is the strongest part with the form of the machine being more or less built around it; at this location the wings are attached and the landing gear fitted, so the region is subjected to all flight and ground loads.

During normal operational flying aircraft are exposed to a variety of stresses: extremes of weather, large temperature changes – from +40°C on the ground to –70°C in the air – cabin pressurisation, wing flexing, landing and take-off loads and engine vibration. The aluminium-copper alloys mostly used in aircraft construction are light and flexible yet strong enough to cope with such conditions. Heavy landings, violent weather or severe turbulence can also be handled by modern jets with safety, although airframe checks may subsequently be required. Normal maximum manoeuvring loads in flighty for modern jets are +2.5 g to –1.0 g, where 'g' is the force of gravity and negative 'g' the impression of weightlessness.

On completion of a new aircraft design, a full test programme is implemented, with an airframe being tested to destruction in a test rig. The wings are bent upwards by large jacks to test strength, and to witness the wingtips almost touching before failure occurs would dispel any fear of flying. On occasions, admittedly inadvertently, aircraft are put to the test in earnest during flight. Fortunately, in the few incidents recorded, the airliners involved have survived and the events have borne witness to the impressive strength of modern machines. One such incident occurred to a Boeing 747 SP (special performance) on 18 February 1985.

China Airlines Flight 006 was en route from Taipei, the capital of Taiwan, to Los Angeles, California, with the aircraft under the command of Captain Ming Yuan Ho. The B747 SP was shorter than the standard 747 model and was specifically constructed for such long-range flights, being designed to fly at high altitudes to conserve fuel. At just after 10:00 local time, Flight 006 was cruising at 41,000 ft (12,500 m) over the Pacific, with the aircraft lying 300 nm north-west of San Francisco. One and a half hours' flying time remained to Los Angeles. The journey so far had

China Airlines 747. *(John Stroud)*

been routine and the 254 passengers had just finished breakfast. Extra crew members were carried for the long duty day, and the five flight deck members and fifteen flight attendants brought the total on board to 274.

In the rarefied atmosphere at 41,000 ft and at a weight of 245 tons, the 747 SP's margin between the maximum and minimum flying speeds was narrow, and could have been as small as thirty knots. At too fast speeds, high speed buffeting caused by shock waves on the wings occurs and the aircraft can be damaged. At too slow speeds, low speed buffeting caused by turbulent airflow over the wings occurs and lift is lost from the wings. If the speed is allowed to decay further the wings can stall and the aircraft can drop from the sky.

The China Airlines 747's autopilot was engaged and the autothrottle of the Performance Management System (PMS) was selected to control the speed at Mach 0.85 (i.e. 85 per cent of the speed of sound at that level). Flying conditions in the daylight were clear, with cloud 10,000 ft (3,050 m) below covering the ocean, but the changeable wind was causing turbulence and was upsetting the delicate balance of the flight. The speed fluctuated from Mach 0.84 to Mach 0.88 and the autothrottle moved in response in an attempt to stabilise the speed. Suddenly, as the thrust levers retarded automatically to flight idle to reduce excessive speed, number four engine suffered a problem. On achievement of the desired speed numbers one, two and three engines increased to the required setting, but number four engine failed to respond. It remained in what is known as a 'hung condition'.

Before flight, Captain Ming had noted from the technical log that number four engine had a history of ceasing combustion, or flaming out, when decelerating. In the turbulent conditions, with the autothrottle advancing and retarding the thrust levers in an attempt to hold speed, the number four engine had required special attention. If engine flame-out occurred, there would be insufficient power from the remaining engines to maintain speed in the rarefied atmosphere and the delicate balance of the flight would be impaired. Captain Ming would have no choice but to descend into the denser air at a lower level.

The flight engineer attempted to restore power to number four engine but his efforts failed and ninety seconds later the engine flamed out. The speed began to decay and the aircraft yawed and rolled to the right. With no automatic rudder, the autopilot applied full left aileron to compensate for asymmetric thrust. Captain Ming disengaged the PMS and the height lock and eased the nose down in descent using the autopilot control wheel. The actions of the captain, however, were insufficient to contain the

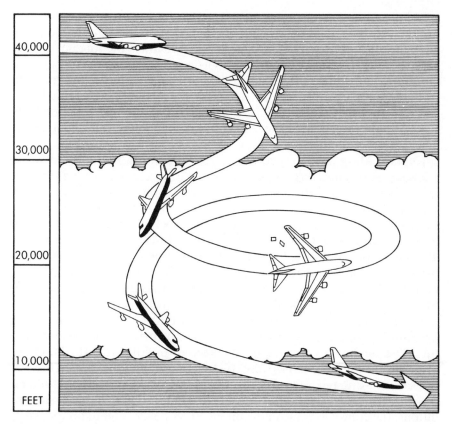

Flight 006's spiral dive.

situation and when Ming disengaged the autopilot to revert to manual control he, and the other 273 occupants aboard Flight 006, received the shock of their lives. Instantly the control wheel snapped to the central position and the aircraft bank increased abruptly 63° to the right, with the speed dropping to Mach 0.75. The big jet began a steep dive towards the ground. Flight 006's nose pitched down to 67° and the aircraft rolled on its back to a bank angle of 160°. The 747 plummeted earthwards at 15,000 ft (4,570 m) per minute and entered the cloud layer which lay below. The crew were caught completely by surprise and in seconds were totally disorientated.

In the cabin all hell broke loose. 'Dishes crashed against the walls and floors', recounted William Peacock, a Vietnam veteran and colonel in the Marine Corps Reserve. 'Baggage compartments opened up and window shades were forced down by vibration.' 'You could feel bits popping off', commented another passenger. Those people not in their seats were thrown about the cabin and fifty passengers received minor injuries. Two flight attendants suffered serious back damage. The aircraft continued its spiralling plunge, screaming almost vertically towards the earth as it dropped at three miles per minute. As the 747 spun through the cloud the airframe and wings were subjected to 5 g, five times the force of gravity and twice the structure's design limit. The aircraft shook violently and the vibrations caused large portions of the horizontal stabiliser to break off. Both outboard elevator sections detached, the auxiliary power unit broke free and the right wingtip high frequency radio antenna separated. By rights the aircraft should have broken apart completely, but somehow the machine held together.

Captain Ming tried desperately to regain control but, with no visual reference and pinned to his seat with the 'g' forces, he faced a difficult task. There seemed little chance of recovery as the aircraft shuddered in a steep diving turn, when, to complicate matters, the landing gear suddenly deployed. The powerful uplatches locking the undercarriage in the bays had released under the excessive 'g' forces and the landing gear had fallen down into place. The two body landing gear doors were ripped from their hinges and flew off into space. The lowered undercarriage, however, had an unexpectedly beneficial effect, as the wheels biting into the airflow seemed to stabilise the aircraft. This gave Captain Ming the chance he needed and by an extraordinary effort he righted the falling machine. Passing 11,000 ft (3,350 m), with only forty seconds to go before the 747 plunged into the sea, he managed to pull the aircraft out of the dive. The three good engines were still functioning and the captain applied power. As Ming regained control, Flight 006 broke from the cloud and the aircraft was levelled off at 9,500 ft (2,895 m). Flying in the clear conditions, the crew were able to regain their composure and to take stock of the situation. Quickly Captain Ming spoke to the passengers and in a 'pretty shaky' voice instructed the cabin occupants to fasten their seat belts.

The captain made the decision to land as soon as possible to inspect the damage and Flight 006 was cleared to divert to San Francisco, the nearest suitable airport. Ming climbed the aircraft back to 27,000 ft (8,230 m) and one hour later, without further incident, the 747 commenced its final

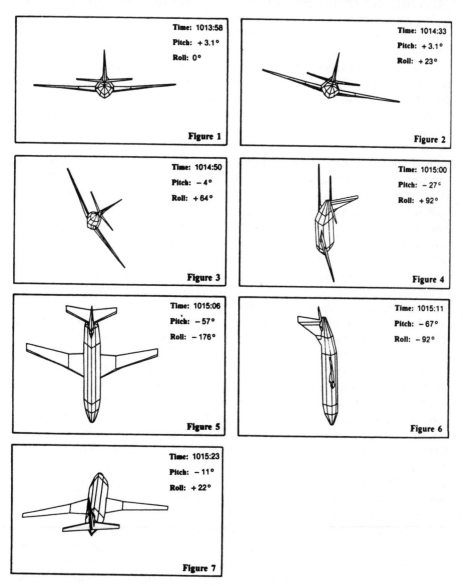

Computer-animated sequence of the 747's spiral dive. (NTSB)

approach. Fortunately the crew were able to lock the landing gear down and to obtain a safe display on the indicators. Flight 006 landed safely and on touchdown Captain Ming and his crew received a spontaneous round of applause.

Mystery surrounded the incident for some months until a subsequent inquiry verified the facts. It was concluded that the actions of the captain when dealing with the engine failure in the delicate circumstances of the high-altitude cruise had been on the tardy side. As a result the autopilot inputs required to maintain control had been excessive. When Ming disconnected the autopilot he was caught completely off guard by the out-of-balance forces, and the machine had spun from his grasp. In defence of the captain, the International Federation of Airline Pilots (IFALPA) stated that too often on long-distance flights aircraft commanders are compelled to fly at very high altitudes to lower fuel consumption in order to complete their journeys. Captain Ming had also done a fine job in regaining control of the stricken aircraft, albeit with some luck from the falling landing gear, and he deserved his applause on landing. The passengers, perhaps, should also have clapped the Boeing Aircraft Company, for the 747 was the real hero of the event. The aircraft was subjected to probably the greatest 'g' forces sustained by any wide-bodied civilian airliner, and survived. No clearer demonstration of the strength of modern aircraft could have been presented and the successful outcome was a fine testament to the skill of the aircraft designers.

Unfortunately the gross jet upset suffered by the China Airlines Boeing 747 was not the only such event recorded in aviation history in the last few decades. In 1979, another Boeing machine, in this case a 727, was involved in a similar upset. The mystery surrounding that incident was never properly resolved and to this day the repercussions of the episode still affect those involved.

On 4 April 1979, TWA Flight 841, a Boeing 727-100, registered *N840TW*, prepared for departure from John F. Kennedy (JFK), New York, for a scheduled service to Minneapolis/St Paul International Airport in Minnesota. In command of the aircraft was Captain Harvey 'Hoot' Gibson. His co-pilot was First Officer Jess Kennedy and at the flight engineer's station was Second Officer Gary Banks. The crew was a very experienced trio, with Captain Gibson having amassed a grand total of 15,710 flying hours at the relatively young age of forty-four. He had learned to fly at only thirteen years of age and after a period serving as an air traffic controller in Chicago had joined TWA in 1963. Gibson was not only qualified to fly the 727 but also had his licence endorsed to fly the DC-9, L1011 TriStar and the Boeing 747. He was one of aviation's 'all-rounders' and he flew helicopters and piloted balloons as well. Captain Gibson had not flown the B727 regularly for some time, however, for he had been operating as co-pilot on the B747 for just over one year. In December 1978 he had broken his ankle and the injury had kept him off work for three months. He had only just returned to his duties as captain on the B727 three weeks earlier and had completed ground school refresher, simulator and aircraft handling details. On 28 March he successfully completed a route flight check of over five hours' duration.

First Officer Kennedy was forty years of age and had joined TWA in 1967. He had operated as a flight engineer for eight years and in 1978 had transferred to the right-hand seat as co-pilot. He had a total of 10,336 flying hours, of which an impressive 8,348 hours had been acquired on the B727. Second Officer Banks was thirty-seven and had operated as a flight engineer for nine years. He was also a qualified commercial pilot. In the cabin were four flight attendants providing for the eighty-two passengers, bringing the total on board to eighty-nine.

The trip for the flight crew began in Los Angeles, California, the day before, on 3 April, and it was Captain Gibson's first time in command without supervision after his spell as first officer on the 747 and his sick leave. Duty had commenced at about 08:30 local time (11:30 Eastern Standard Time, now used throughout) and after a series of short sectors had terminated in Colombus, Ohio, at 22:00. After a night's rest the crew had recommenced duty the following day, Wednesday 4 April, at 13:45, and in the afternoon had operated Flight 841 via Philadelphia, Pennsylvania to JFK, New York. The B727 had arrived at Kennedy in the evening at 17:20. With only eighty-two passengers scheduled for the next sector to Minneapolis/St Paul, a comfortable turnround should have been accomplished, but it was not to be. New York in the 1970s suffered from a surfeit of aircraft movements, and the mid-week traffic congestion was causing extensive delays.

Night fell as Flight 841 prepared for departure, and while waiting for take-off a score of anti-collision lights could be seen twinkling ahead in the darkness. After forty-five minutes of taxiing, three-quarters of a ton of fuel had been consumed from the sixteen-ton load on board, even with one engine shut down to conserve fuel. Finally the B727 was given clearance for take-off, and at 20:25 Flight 841, weighing sixty-five tons, lifted off from JFK. The aircraft climbed steadily in the blackness to its planned cruising height of 35,000 ft (10,670 m) and twenty-nine minutes later, at 20:54, levelled in the cruise. The cockpit crew settled down to their usual routine and all was normal except for an exceptionally strong headwind.

The scheduled flight time for the 900 nm journey was just over two hours, but the very high winds being encountered slowed their progress. The ground speed was down to about 380 knots and at this rate the flight time would be nearer two and a half hours. The B727 would be further delayed and additional fuel would be consumed from the load already eroded by the waiting time at Kennedy. Captain Gibson hoped the winds might slacken later in the flight and in the meantime the crew ate their evening meal.

Flight 841's routeing crossed the border into Canada over the northern end of Lake Erie, then passed over London, Ontario, before re-entering the States over Michigan. About one hour after departure the 727 entered Canadian airspace and First Officer Kennedy established contact with Toronto Control. Captain Gibson was still concerned about the very strong headwind which blew in excess of 100 knots and, with the aircraft now light enough to climb, he felt it might be prudent to change altitude to 39,000 ft. The decision, however, was not taken lightly. The 727-100 was not considered by pilots to be a comfortable aircraft to fly at such a high altitude, but it was felt the slacker winds expected at the upper level would make the climb worthwhile.

'Toronto Centre,' radioed Kennedy, 'have you any reports of winds from flights at other levels?'

'Negative,' replied Toronto, 'no reports.'

'We're encountering headwinds in excess of 100 knots and we'd like to climb to 390.'

'Roger, TWA 841, climb and maintain level 390.'

Captain Gibson disengaged the autopilot and initiated a manual cruise climb, gradually ascending the aircraft to the upper level. At 21:38, 39,000 ft (11,890 m) was achieved, and the autopilot and altitude hold were reselected. Second Officer Banks adjusted the thrust levers to maintain Mach 0.81. The winds at the new level were down to 85 knots, which helped improve the ground speed, and the fuel consumption was better. The outside air temperature indicated –57°C. The flight engineer estimated that about six and a half tons of fuel had been used and at 21:40 he noted the aircraft weight of fifty-eight and a half tons in the flight data log. The night was clear with cloud cover far below and a half moon shone about 50° above the horizon. The ride was smooth and the climb to 39,000 ft seemed justified. First Officer Kennedy set about recalculating the aircraft's ground speed and adjusting the leg times in the flight log while the captain monitored progress.

Flight 841 crossed the Canadian/US border into Michigan State and continued north-west between the city of Detroit to the south and Lake Huron to the north. The city of Saginaw, situated just south of Saginaw Bay, lay ahead. The sky was clear and bright, almost like daylight with the brilliant, near-full moon, and a thin but solid overcast lay directly below at 35,000 ft (10,670 m). Captain Gibson began sorting his charts for the onward route and he turned briefly to his left to extract fresh maps from his flight bag which lay by his left side on the cockpit floor. The time was 21:47. As his attention was momentarily distracted he suddenly felt a strange buzzing sensation. Within seconds the buzzing increased to a light buffet and he could feel the effect through the airframe.

Quickly Gibson looked at the instruments and checked his visual reference in the clear night. The aircraft wings were level but the autopilot had turned the control wheel 20°–30° to the left. The situation seemed very strange indeed. The other two on the flight deck were still busy with their calculations and at first had not noticed the development. Captain Gibson disconnected the autopilot and maintained the wings level by holding the control wheel to the left. Suddenly, the aircraft yawed sharply to the right, followed by another severe right yaw. All on the flight deck were surprised by the manoeuvre. The second yaw continued, with the aircraft rolling and skidding in a turn to the right. The nose dropped swiftly and a rapid roll to the right developed. The captain gradually pulled back on the control column to the limit with a slight backwards pressure and applied full left aileron. Neither input was sufficient to halt the manoeuvre. As the aircraft banked through 30°–45° to the right, Gibson applied full left rudder. For a moment the roll rate slowed, then the rapid right roll continued. The sky was crystal clear in the bright moonlight and, with the situation deteriorating so swiftly, all control inputs and decisions were made with reference to the visual horizon. Gibson managed a last glimpse of the artificial horizon as the machine banked through 45°–60° to the right and felt the aircraft was about to turn over. In a desperate effort to save the situation he took his right hand from the control column and quickly retarded the thrust levers to flight idle. His efforts, unfortunately, were to no avail. The aircraft continued to roll rapidly in a right diving turn, with the nose dropping sharply, and there seemed nothing the captain could do.

'We're going over', he shouted.

The 727 flipped over on its back and, with the nose below the horizon, plunged earthwards close to the city of Saginaw. The engines hiccuped with the disturbed airflow and two quick, loud bangs of the compressors

TWA Boeing 727. *(John Stroud)*

stalling could be heard in the cabin. For less than a second a feeling of absolute terror gripped Captain Gibson and then, strangely, a near calm returned almost as quickly. It was as if he knew that in spite of any recovery attempt his efforts would not be sufficient and that it was all over. Suddenly he felt relatively cool and his mind was clear. The rapid roll rate continued and by the end of the first roll, as the wings passed through level again for the first time, the nose had dropped 30°–40° below the horizon.

'Get them up', shouted Gibson, referring to the speed brakes.

Kennedy did not understand the command so the captain pulled the speed brake lever himself. The operation seemed to have no effect so he recycled the speed brakes to check the function. The fast roll rate continued unabated and by only one and a half rolls the nose was pointing nearly vertically downwards towards the cloud layer below. What had begun as a barrel roll had quickly developed into a vertical, screaming dive. The aircraft was now pulling 3.5 g or more (727s are stressed to +2.5 g or –1.0 g), so the machine was already being strained beyond design limit. The 727 continued to race earthwards almost vertically with the wings rotating at about 50° per second. Gibson fought desperately to regain control. As the aircraft broke through the cloud layer at 35,000 ft (10,670 m) the captain could see, framed in his windscreen, the ground spinning directly below. It was not a panoramic sweep of the horizon as experienced in a flat spin in a light aircraft, but a vertical, rotating dive. Gibson could see a large dark patch, possibly a forest or lake, and one large city and three small cities spinning before his eyes.

The 727 plunged towards the ground at an average descent speed of 46,000 ft (14,020 m) per minute, with the rate of descent at moments reaching 76,000 ft (23,165 m) per minute. At such speed the aircraft broke the sound barrier and sonic booms could be heard clearly on the ground. On board, the noise level was painful to the ears. Still the 727 raced downwards in the darkness, rolling continuously, and the aircraft buffeted and shook with the fast speed. The artificial horizon which, in level flight, normally showed the top half sky blue and the bottom half earth black, displayed only a solid black. By 30,000 ft (9,145 m), the airspeed indicated about 450 knots on the instruments and the altimeters ran down so quickly the numbers were blurred and were difficult to read. The situation appeared hopeless and all Captain Gibson's efforts seemed of no effect. Still the machine dived and rolled inexorably towards its destruction. The spinning 727 still subjected the airframe to an increasing gravity force and in the cabin passengers were pinned to their seats. Those people standing at the beginning of the manoeuvre were forced to the floor. No one could move and fortunately only minor injuries were sustained. Many passengers found difficulty in breathing and some passed out for a few seconds under the strain.

On the flight deck Gibson watched the airspeed race past the limit and reach 470 knots in the dive. The captain estimated the aircraft to be descending through 20,000 ft (6,095 m), still out of control after about five to six rolls, and drastic action was needed. Quickly he called for extension of the landing gear. It was not permitted to raise or lower the undercarriage above 270 knots because of risk of damage to the gear doors, but in this situation that was hardly a consideration. Kennedy immediately selected the lever to the down position and the crew heard what sounded like an explosion as the landing gear dropped into the excessive airflow. The noise was unbelievable and for a moment Gibson thought the wings had separated. Both main landing gear doors and operating mechanisms were severely damaged and hydraulic line 'A' was ruptured. The right landing gear nearly came off, being blown past the over-centre position. It twisted rearwards into the trailing edge flap track and canoe fairings, jamming the flaps. The lowering of the undercarriage, however, appeared to have the desired effect and the speed at last began to reduce. The 'g' force decreased and the captain noticed some response return to the control column. At the completion of the last spin he managed to stop the aircraft rolling in just over one second, but as he fought to gain recovery the wings continued to rock about 20° to either side with little control. The nose, however, still pointed downwards almost vertically and with the height passing through 11,000 ft (3,350 m) and the aircraft still dropping rapidly, swift action was required.

Maintaining the wings level, he heaved steadily back on the controls and desperately tried to pull the aircraft from its dive. The 'g' forces quickly increased as the 727 began to respond and at about 8,000 ft (2,440 m) the captain appeared to have regained control. *N840TW*, however, was still descending and it was another 3,000 ft (914 m) before it bottomed out of the dive. At 5,000 ft (1,525 m) Gibson continued to pull back on the control column and he could feel his stomach drop to his knees as the aircraft was subjected to 6 g. The wings were tearing at their roots as they withstood six times the aircraft's normal weight. Miraculously they held. The Boeing jet had dropped 34,000 ft (10,360 m) in forty-four seconds, an average descent rate of 46,000 ft (14,020 m) per minute, or over one mile about every eight seconds.

Gibson had pulled the nose from near vertical to arrest the descent, but before the captain could react it passed through the level position and the aircraft shot skywards again. The nose pitched up to about 50° above the horizon and the speed dropped rapidly. Captain Gibson was so intent on not hitting the ground he nearly looped the plane. He had to work hard to prevent the machine from stalling out of the sky and plummeting earthwards once more. Climbing through 8,500 ft (2,590 m) the airspeed decreased to 280 knots and by 11,000 ft (3,350 m) it was down to as low as 160 knots. Using the moon as reference and with Kennedy and Banks both calling out pitch attitudes from the artificial horizon, Captain Gibson managed to regain his bearings and gradually he eased the nose forward. The 727 descended below 10,000 ft (3,050 m) again but, with the speed increasing to 180 knots, he was able to ease the stricken machine back into the climb. By 13,000 ft (3,960 m) he managed to level the aircraft with guidance from the other two and he was able, finally, to make a full recovery. It was an amazing feat of airmanship.

Under the circumstances there was no choice but to land as soon as possible and the captain informed the passengers that everything was now under control and that they would be making an emergency landing at Detroit. By now the flight engineer had taken action over the warning light indicating failure of hydraulic system 'A' and the flap extension drills were commenced. The trailing edge flaps were jammed by the twisted right landing gear and would not move and the leading edge flaps could only be extended by the alternative system.

On completion of the leading edge slats extension, Gibson once more experienced control difficulties. At speeds below 200 knots and above 220 knots the aircraft rolled uncontrollably, this time rapidly to the left, and it was necessary to fly at about 210 knots to maintain control. Attempts to retract the leading edge slats failed. Only inboard ailerons were available, compounding the problem, for the outboard ailerons, which normally come into play with the selection of trailing edge flaps, were still locked in with the flaps being jammed. Part of the rudder was also inoperative with 'A' system hydraulic failure.

The aircraft buffeted severely, making it impossible to read some of the instruments, and Gibson was unable to verify his height. In the distance the crew could see the lights of Chicago, where the captain had once been an air traffic controller, but it was feared too far for a landing. The aircraft might not hold together till then, so the nearer Detroit Airport had been chosen, although the weather there was much worse with a low cloud base. The damaged landing gear also displayed an unsafe condition with three red warning lights indicating all three gears unlocked. Nothing could be done for the two main landing gears, but a manual wind-down facility was available on the flight deck for extending and locking the nose gear. Second Officer Banks began to crank the mechanism. As the three cockpit crew performed their emergency checklist procedures, the flight attendants in the cabin prepared the 82 passengers for an emergency landing.

After about thirty minutes, approaching Detroit, Banks succeeded in locking down the nose gear and stowing the doors. The nose-gear light now glowed green and the stowed nose-gear doors reduced the buffeting by about forty per cent, which at last enabled the captain to read all the flight instruments.

The aircraft was proving very difficult to handle and Captain Gibson gingerly commenced an approach to Detroit's 03L runway, fighting the aircraft down through the cloud. He did not intend to land from this first approach but planned to execute a low-level pass over the airport to allow an examination of the landing gear. Over 90° of aileron control wheel input was required, with the rudder held fully to the right, to maintain control. Even with maximum right trim selected, the strain on Gibson's right leg was enormous and he could feel his leg pulsating to the beat of his heart. The weather did not help, either, with freezing drizzle falling from a cloud base only a few hundred feet above the ground. The 727 broke from the cloud at about 400 feet and Gibson flew down the runway on a low-altitude pass at about 50–75 ft (15–23 m) while crash rescue personnel shone searchlights on the landing gear. The gears were checked in place by the tower controllers but much damage could be observed and the right gear appeared misplaced.

On being so close to the runway, Captain Gibson had the greatest urge to jump from the aircraft and place his feet firmly on solid ground rather than go back up into the weather. The captain commenced a left circuit of the airport, staying low and trying to keep in visual contact with the runway. Turning to the left at the start of the downwind leg he concentrated hard on maintaining sight of the lights as he executed the uncomfortable manoeuvre. The aileron control was still held almost fully to the right, with hard right rudder, and his right leg still pulsated with the strain. Suddenly the captain lost control and he was unable to level the wings. The left turn continued and the 727 headed back towards the airport. The co-pilot also grabbed the control wheel, holding on full right

Captain 'Hoot' Gibson at the controls of a Ford Trimotor.

aileron, but was unable to stop the turn. Gibson pushed the number one thrust lever to high power and retarded the centre and right engines to idle. Gradually the asymmetric power took effect and, just before crossing the runway, he managed to gain control and to bring the wings level.

The next approach, this time for a landing, was flown at about 205 knots and the 03L threshold was crossed at the same fast speed. Gibson flared the aircraft with full right rudder and aileron applied and at 22:31 executed a beautifully smooth landing. The 727 landed with the left gear touching down first at a speed of 197 knots. Fortunately it held. On closing the throttles, however, the machine rolled uncontrollably to the left. Reverse thrust was selected as normal and the nose wheel was lowered to the ground. As yet the right gear had not touched and the crew feared it had gone. As the 727 edged towards the left side of the runway the captain slowly lowered the right wing and at a considerable angle the right main landing gear finally touched. By good chance it also held and the aircraft continued its rollout down the runway with a pronounced right wing low attitude. Sparks showered from the dragging right landing gear and the tower controller radioed advising of a possible fire. Nose-wheel steering was not available with the loss of hydraulic system 'A' and Gibson had to steer the aircraft off the runway along a high-speed turn-off using differential braking and asymmetric reverse power. On the taxiway, the 727 was met head-on by fire trucks racing towards the stricken aircraft, and Gibson finally brought the machine to a halt. It was a fine demonstration of flying skills in very trying circumstances.

Foam was sprayed on the damaged areas and one of the firemen on the ground called the captain on intercom to advise that 'fuel was running all over' and that an emergency evacuation using escape slides might be prudent. Captain Gibson felt that his passengers had been through enough and that with fuel and fire trucks all over the place in the dark-ness it would be safer to walk the passengers off the aft exit. He replied accordingly. The plane was only about half full anyway and it would not take long to disembark the passengers from the rear. Quickly all on board evacuated by the aft steps and a relieved group of passengers and crew were finally taken to the terminal. They had faced death and survived.

An attempt was made later to tow the 727 from the high-speed turn-off but when the aircraft had been moved less than ten feet the right landing gear started to separate. The machine was jacked up to examine the damage but as the wing was raised the right landing gear broke from the wheel well and fell in three separate pieces.

A further inspection of the aircraft the next day revealed extensive damage. Wing skin panels were buckled, bolts had been sheared and large pieces of the machine had virtually been torn off. Wrinkling of fuselage panels was evident at the wing roots and clearly indicated the strain to which the wings had been subjected. Fuel leaked from around several structural fasteners in the left wing. A speed brake section had been wrenched from its hinges, an aileron bolt had been severed, a trailing edge flap carriage had been damaged and a flap transmission mechanism had been broken. All landing gear doors were extensively damaged and hydraulic system line 'A' was fractured. A flap track fairing was missing, as were some engine panels, and a leading edge lift device had sustained damage. To be precise, the number seven leading edge slat mechanism had suffered sheared bolts and twisted tracks and more significantly the slat itself was missing. Could this be the cause of the spiral dive of the 727? A search of the Saginaw area was conducted for the missing aircraft parts and the number seven slat, broken in two, the flap track fairing, and most of the speed brake section were found about six to seven miles (ten kilometres) north of the city. The stricken 727 lay at the side of the runway while the landing gear was made safe and was eventually towed to the hangar for repairs. Several days after the incident the aircraft was sufficiently repaired to permit a ferry flight to the TWA maintenance base at Kansas City, where major repair work was conducted.

Captain Gibson and his crew received just praise for their skills in saving the stricken 727 and the captain received a letter of commendation from the Federal Aviation Administration (FAA) via TWA's vice-president of operations. The cause of the incident was unknown but both the FAA and TWA exonerated the crew from any blame.

Immediately the National Transport Safety Board (NTSB) began an investigation of the accident, including the part played by the crew in the upset. As a formality, Captain Gibson was required by the FAA to be re-rated in the 727 simulator, as well as on the aircraft, and to attend a high-altitude awareness course. He passed with flying colours and TWA received a further letter from the FAA complimenting the captain on his abilities. Ten days after the incident Gibson was promoted to Lockheed L-1011 captain.

The investigation of the upset proved to be a lengthy affair and it was to be more than two years before the NTSB reached a conclusion. Fortunately the B727 involved in the incident was not compromised by the delay and, although damaged extensively, *N840TW* was repaired within a matter of weeks. It was duly returned to service, none the worse for its ordeal, in late May 1979. Sadly, the same could not be said for the flight crew. As the inquiry continued, the three men found the situation changing for the worse. Adverse rumours began to circulate that more was involved than met the eye and Gibson and his colleagues were placed unfairly under suspicion, as much by the piloting fraternity as anyone else. The trio faced a harrowing ordeal and the mental anguish suffered made their return to duties and a normal life substantially more difficult.

The initial suspicion that the number seven slat was responsible for the gross upset prevailed, but how and why the slat had extended whilst

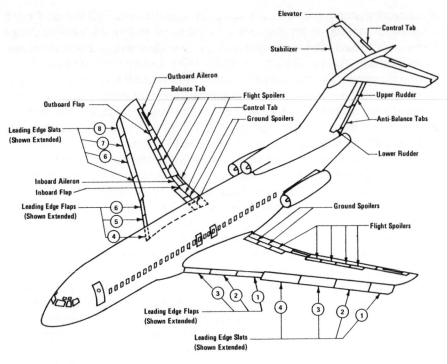

Boeing 727 flight control system surfaces. (NTSB)

cruising at 39,000 ft (11,890 m) remained a mystery. Lift devices on the leading and trailing edges of the wings of modern jet aircraft are essential for take-off and landing but are not designed for use at such high altitudes. On modern jets the wings are swept back at a large angle (on the B727, 34°) to allow the aircraft to fly high and fast by delaying the onset of shock waves as airflow over the wings approaches the speed of sound. At slow aircraft speeds, however, the lift-producing qualities of the wings are poor. To improve lift, high-lift-producing devices in the form of slats (see diagram) are required at the wing leading edges*, and flaps are required at the wing trailing edges. When extended these devices increase the wing surface area and the camber of the wing shape. With slats and flaps fully extended the wing area is increased by twenty per cent and the lift by over eighty per cent. To improve lift at take-off, some flap and all slats are extended, any increase in drag being more than compensated for by increase in lift, and take-off without slats and flap is not possible at

---

* The Boeing 727 also has a type of leading edge flap, but these are not relevant to the story.

normal operating weights. On landing, slats and full flap are always selected in normal circumstances. Leading edge slats and trailing edge flaps are set together by the operation of a single lever on the flight deck. With selection of 2° of trailing edge flap the numbers two, three, six and seven slats automatically extend, and with selection of 5° of trailing edge flap the remaining slats, numbers one, four, five and eight also automatically extend. With 5° of trailing edge flap set, all the leading edge devices are extended, and further operation of the flap lever selects only additional trailing edge flap extensions until full landing flap is set.

At cruise speeds and levels there is no requirement to select slats or flaps, and operation of these lift devices on the B727 is expressly forbidden above an indicated airspeed of 230 knots and an altitude of 20,000 ft (7,620 m). Rumour later had it, however, that flaps on the B727 were being extended by crews at higher altitudes than permitted to improve performance, and that the habit was not just confined to a few people but was fairly widespread. Since the use of leading edge devices was dangerous at the higher flight levels and speeds, 727 crews, it was alleged, were pulling the electrical circuit breaker which shuts off the hydraulic valve feeding the leading edge slats, thereby isolating the mechanisms in the retracted position, and then extending the trailing edge flap to 2° by use of the flap lever. The hydraulic valve cut-off circuit breaker was situated beside the crew wardrobe, on the right of the cockpit in a recess aft of the flight engineer's panel.

The effect of the small amount of trailing edge flap in the cruise, it was rumoured, tucked the nose down and improved speed and fuel consumption. The implications were, therefore, that the pilot of Flight 841 had deliberately selected the trailing edge flaps and that by some error the number seven slat had extended and become isolated in that position. In a landing configuration, an extended number seven slat would provide extra lift and would raise the right wing, rolling the aircraft to the left. In a fast, high altitude cruise with a low nose attitude, however, the device would create the opposite effect. When extended the number seven slat would produce negative lift on the right wing and the aircraft would roll sharply to the right.

Flight and simulator tests subsequently appeared to demonstrate that extension of the number seven slat had caused the incident and that it had detached near the end of the last vertical roll, at about the time of lowering the landing gear, and had allowed Captain Gibson to regain control and save the aircraft. Whether the extension was an unscheduled movement or, as whispered, an irresponsible crew action, however, remained unresolved. No proof of any crew involvement was forthcoming although the stigma remained. Rumours implicating the pilots in the event continued unabated and it appeared, on hearsay alone, that a crew who should have been treated as heroes were being seriously maligned. The crew protested that they had never even heard of such a malpractice, but it was to no avail. At the time of Gibson's incident few, if any, knew of the action of selecting flaps in the cruise, but afterwards rumour had it that the practice was widespread. Suddenly everyone seemed to know someone who had tried it although no one who had actually used the procedure emerged. In answer to the accusations of tampering with the flap controls, Captain Gibson, in a deposition to the inquiry in Los Angeles, California on 12 April 1979, one week after the incident, issued a sworn affidavit.

'At no time prior to the incident did I take any action within the cockpit either intentionally or inadvertently, that would have caused the extension of the leading edge slats or trailing edge flaps. Nor did I observe any other crew member take any action within the cockpit, either intentional or inadvertent, which would have caused the extension.'

The first and second officers also both issued sworn statements at the same place and time denying tampering with the flaps. The deposition was conducted in the full glare of publicity with TV cameras and a battery of twenty-six microphones.

These testimonies seemed sufficiently clear but they did not satisfy the NTSB. A statement, allegedly made by one of the investigators a few months later, clearly laid blame on the crew's shoulders. 'I think those guys were fooling around up there and I don't think we really know what they were doing yet', a 'spokesman' revealed to *Aviation Consumer* magazine on 15 October 1979. Such a statement, if accurately reported, was a disgraceful breach of ethics, for the investigation was still at a premature stage. Soon the national press echoed the comment and a countrywide accusation of guilt resulted. The crew seemed convicted before any hearing had begun. Meanwhile, Captain Gibson and his colleagues, desperately trying to pick up the pieces of their careers, were restrained from issuing public statements in their own defence.

A number of points surrounding the incident were also found questionable by members of the inquiry. To begin with, the cockpit voice recorder (CVR) tape on the 727 had been erased after Flight 841's emergency landing at Detroit. Cockpit voice recorders are installed on aircraft to record flight deck conversations which might be useful to investigators following an accident. CVRs are not permitted to be used by the FAA in any disciplinary action. The recorder tape operates on a continuous, thirty minute loop and would not have recorded details of the dive, but the NTSB were suspicious that conversations on the flight deck after the incident could have revealed details of the crews' actions and were deliberately erased. In 1979 the CVR had been only recently introduced and crews distrusted the equipment. Most pilots feared misuse of its information and regularly erased the tape after a trip. Under questioning by counsel, Captain Gibson admitted that he, too, normally erased the tape after a flight but that in this instance he could not remember doing so.

'Did you erase the recorder?'

Captain Gibson: 'Not to my knowledge.'

'Did anyone erase it?'

Gibson: 'Not to my knowledge. I didn't see anyone erase it.'

'Do you usually erase the recorder?'

Gibson: 'I usually do, yes. I don't recall erasing it.'

'Can you erase it in the air?'

'Gibson: 'No.'

'What is required to erase the CVR?'

Gibson: 'The parking brakes have to be set.'

'How many minutes of recording are there on the CVR before previous contents are erased?'

Gibson: 'Thirty minutes. It was forty-five minutes to an hour [after the incident] before the aircraft was shut down and we got off.'

'So if the tape had not been bulk-erased at the time of shut-down would there have been anything meaningful on the tape?'

Gibson: 'No, the tape could only have made my other two crew members look good. They did a real good job. All that would have been on the tape would have been the other crew members complying with the check list.'

'Can you explain why it is your habit and routine to erase cockpit voice recorder data on landing?'

Gibson: 'It is an accepted practice, and as far as I am concerned at the time it was done by everyone. It is done by an awful lot of people. When they put the cockpit voice recorder on the airplane I would say 100 per cent of people always erase it on landing after they park their brakes.'

'Why do you do it?'

Gibson: 'Because I might say something unkind about some of the people in management, and they might take that tape out and send it someplace.'

Both Kennedy and Banks also swore that they had not erased the tape. It seemed reasonable that, in the highly charged atmosphere after such an event, crew members might by force of habit take actions which they might later forget, but the investigators were not to be convinced.

'We believe the captain's erasure of the CVR is a factor we cannot ignore and cannot sanction. Although we recognise that habits can cause actions not desired or intended by the actor, we have difficulty accepting the fact that the captain's putative habit of routinely erasing the CVR after each flight was not restrainable after a flight in which disaster was only narrowly averted. Our scepticism persists, even though the CVR would not have contained any contemporaneous information about the events that immediately preceded the loss of control, because we believe it probable that the twenty-five minutes or more of recording which preceded the landing at Detroit could have provided clues about causal factors and might have served to refresh the flight crew's memories about the whole matter.'

The investigating team clearly felt that the erasure of the CVR was deliberate. If the inadvertent operation of the number seven leading edge slat was also to be proved part of a deliberate act by the crew, then any mechanical failure of components would have to be discounted. The NTSB instigated a thorough investigation of the possibilities of an unscheduled extension of the slat. If some kind of mechanical fault had resulted in accidental operation of the number seven slat, the Safety Board would do their utmost to find the cause. A total of 118 trials were conducted in a Boeing flight simulator to try to identify the condition that precipitated the aircraft's upset and to duplicate and evaluate its manoeuvre.

'The flight simulator traces showed that the simulated aircraft could be returned to wings-level flight with relatively little loss of altitude provided corrective action was begun before the roll and airspeed were allowed to increase excessively. In the simulations, the pilot could delay reaction for about sixteen seconds and regain control with an altitude loss of about 6,000 ft (1,830 m). However, when the pilot delayed corrective action for seventeen seconds or more, a manoeuvre was entered that approximated Flight 841's airspeed, altitude, and g-traces. In this manoeuvre, the aircraft continued throughout the descent to roll to the right, in spite of full left aileron and rudder, until the slat was retracted to simulate its loss from the aircraft.'

During none of these tests, however, was yaw induced at the beginning of the upset.

A thorough inspection of the damaged aircraft and the retrieved debris was conducted but no evidence of a malfunction which might have caused an inadvertent slat extension was discovered. An examination of the slat locking devices also proved fruitless. The slats are locked in the retracted position by hydraulic pressure from system 'A' and by a mechanical locking mechanism. Both systems would have had to fail before the slat could have extended and there was no evidence to suggest that this had occurred. Loss of hydraulic system 'A' and/or failure of a mechanical lock would have illuminated warning lights in the cockpit, and neither were reported by the flight crew.

A repair of the hydraulic system line 'A', which ruptured on extension of the landing gear, was implemented, and the slat and flap systems were found to function satisfactorily. Number seven leading edge slat, the offending device, could not be tested since it had separated from the wing, and its extending and retracting hydraulic mechanism, or actuating system (see diagram), could not be checked since it was broken. The investigating Board acknowledged, however, that failure of the actuating system, namely, fracture of the actuating piston or separation of the piston from the actuating rod, could have resulted in inadvertent extension, but Boeing convinced the NTSB that the safety margins, particularly regarding an actuator piston fracture, were massive and that in sixteen years of service history and over thirty-six million flight hours such a fracture had never occurred.

There is always a first time, of course, but the event was deemed by the Board to be highly unlikely and was not considered. Later it was revealed that at the time of the inquiry no fewer than six cases of cracked actuator pistons had been reported to Boeing by Lufthansa. Separation of the actuator rod from the actuator piston seemed to the Board to be the only reasonable single failure which could have permitted an unscheduled slat operation, but this was also discounted. The Safety Board contended that the distortion of the remaining section of the broken actuator cylinder and the fact that the piston did not stay in the remnant of the cylinder indicated that the piston rod was attached to the piston when the slat extended. The problem facing the NTSB, however, was that the forward two-thirds of the actuator cylinder, the actuator piston and the actuator piston rod were missing and in their reports the investigators could only speculate as to the reason for the failure. The Safety Board had to reach their conclusions only on the limited evidence available and it seemed a firm dismissal of any mechanical failures, especially the only single failure which was considered reasonable, when the suspect parts were never recovered.

Over the years, a large number of 'service difficulty reports' had been received by the Federal Aviation Administration (FAA) outlining slat problems, most of them minor, but some significant. Whether the 727 involved in the incident, *N840TW*, was the subject of such reports is not

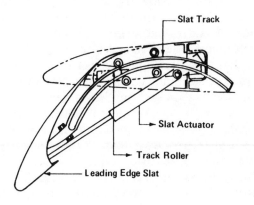

Typical slat section. *(NTSB)*

known, but it was an old aircraft and was TWA's first 727-100 in service. Up to 1973, there had been approximately fifteen reported cases of uncommanded single leading edge slat extensions at low level with the slat separating in flight. To resolve the situation, in 1973 Boeing issued an Airworthiness Directive requiring a stronger actuator piston rod end, and the problem was assumed to be solved. In the five years preceding Gibson's upset, only two instances of importance were reported. One notable event arose in 1976 when an unscheduled extension of a leading edge slat occurred. Significantly, the extension resulted from the failure of a slat actuator support fitting. The slat remained attached and the aircraft upset was contained.

In 1978, one B727 operator declared that in cruise at 25,000 ft (7,620 m) the numbers six and seven leading edge slats had been unintentionally extended by the crew. The captain stated that he believed the trailing edge flaps were partially extended and when he attempted to retract them the leading edge devices were accidentally operated. The captain immediately retracted the leading edge slats, but the numbers six and seven slats remained extended. The roll to the right was contained, and when the aircraft slowed the slats retracted. Whether rumours of flight crew tampering with the flap controls surrounded this upset was not revealed, but the NTSB appeared to be drawing comparisons between this event and Captain Gibson's. Since the investigators had discarded the likelihood of mechanical malfunction, the Safety Board were forced to consider that Flight 841's crew, in spite of their statements, had deliberately operated the flaps.

The NTSB advanced the theory, therefore, that the electrical circuit breaker for the leading edge lift devices had been pulled to prevent their extension and the trailing edge flaps had then been set to 2° to improve aircraft performance. The Board then tried to imagine a scenario in which the number seven slat could have been inadvertently extended and were aided in their quest by what appeared to be an unusual source, Captain Gibson himself. In an unofficial interview given shortly after the incident, an FAA representative had asked Gibson if all the crew members had been in their places when the upset occurred. The captain replied in the positive, adding that the flight engineer had just returned to his seat. The FAA representative assumed from the statement, incorrectly, that Banks had actually left the cockpit and, unknown to Gibson, the NTSB were informed accordingly. The suspicions of the Safety Board were corroborated by a female passenger in economy class who stated that she saw a male crew member carry trays back to the galley from the forward end of the aircraft. The unidentified man was wearing epaulettes and was assumed by her to be one of the flight crew. Later it appeared that one of the stewards, who also wore epaulettes at that time, had carried the cockpit crew's meal trays back from the flight deck. The Board, therefore, attempted in secret to build a scenario around an erroneous premise and surmised that the setting of the flaps might have occurred with Second Officer Banks absent from the flight deck. The relevant circuit breaker was to be found aft of the flight engineer's station and, on his return, it was suggested, he may have found it tripped. Unaware of the circumstances, Banks may have simply pushed it home, resulting in extension of the leading edge devices. The captain would then have selected the flap lever to up but, in something similar to the previous case, the number seven slat remained extended. An examination of the offending slat mechanism seemed to support the theory. A sheared bolt was shown by metallurgical tests to have been weakened by fatigue and had probably broken before the event. The failure would have resulted in sagging of the inboard end of the slat, which seemed to be confirmed by slat wear. The number seven slat would, in this condition,have extended misaligned and aerodynamic loads would have prevented its retraction.

A broken bolt was also found on the outboard aileron's mechanism which had been weakened similarly by fatigue. It was not possible to determine when the bolt had failed but, if it had sheared before the barrel rolls, the free play in the mechanism would have permitted the aileron to float upwards about one inch (twenty-five millimetres). This would have induced a roll to the right which would have been noticeable to the pilot and would not have helped in controlling the gross upset. All the events could, of course, have occurred simultaneously, but the NTSB investigators concluded that, even had the broken aileron bolt existed at the moment of the upset, it would not have contributed to the loss of control of the aircraft. There were those who disagreed. The US Air Line Pilots' Association (ALPA) were also participating in the investigation and they were becoming increasingly disturbed by the trend of the inquiry. ALPA instigated its own study and claimed that in the five years before the incident no fewer than 400 'service difficulty reports' of B727 slat problems had been submitted to the FAA.

'ALPA is concerned that this incident is not limited to a single aircraft on a single airline. Rather it appears to be symptomatic of the fatigue problem that has not been properly addressed by the aircraft industry. Recent emphasis on this ageing problem associated with the older jets throughout the world underlines ALPA's concern.'

The Pilots' Association felt that the large number of slat problems which had been reported tended to support the mechanical failure theory. The Safety Board's suggestion of malpractice, therefore, appeared hasty and their rejection of the only single failure considered reasonable seemed open to argument, especially since the relevant parts were missing. The only previously reported incident of an unscheduled slat extension also related to problems with the actuator, although in that case it was failure of the actuator mount fitting.

'We believe', reported ALPA, 'that pre-existing fatigue, corrosion and component failures within the number seven slat and right outboard aileron mechanisms caused the slat to extend. We further believe that free play in the right outboard aileron played a significant part in the controllability of the aircraft and initiation of the manoeuvre.'

Flight tests were conducted by Boeing in October 1980 to examine the effects of slat extension in conditions as close as practicable to *N840TW*'s incident, but these also resulted in discrepancies. The NTSB noted that the decrease in airspeed which accompanied the selection of numbers two, three, six and seven slats on the test aircraft compared with the airspeed decrease indicated by Flight 841's flight data recorder in the moments preceding the upset. This, the investigator claimed, was evidence that the relevant leading edge devices had originally extended before the number seven slat had become isolated and was conclusive proof of their theory. Captain Gibson's control deflections of elevator, aileron and rudder in an attempt to maintain level flight, however, could, likewise, have caused a similar reduction in speed. There were also significant differences in other areas, namely the 'g' forces at the onset of buffet. The effect of selecting leading edge slats in the test in fact produced more than moderate buffeting, while Captain Gibson had testified that only light buffeting was experienced. The 'g' traces supported the captain's statement. Gibson's colleagues on the flight deck were not at first aware of the developing situation before the spiral dive, a fact which also seemed to be in keeping with the buffeting being light. If the buffeting had been moderate their attention would have been more readily drawn to the circumstances. No passengers aboard *TW 841* reported any shaking of the aircraft prior to the upset. Light buffeting, however, pointed to extension of only the number seven slat, but no test was conducted with the aircraft in that condition.

On 29 January 1980, almost ten months after the upset, the three flight crew were interviewed for the first time by NTSB investigators in Kansas City, Missouri. The line of questioning was directed at the single issue of the location of the second officer at the time of the incident and clearly indicated the train of thought of the investigators. Demands to be questioned on all aspects of the case by Gibson and the FAA were ignored. Captain Gibson and his crew once again gave testimony but, on this occasion, only regarding the position of the flight engineer, and no other depositions were made.

The US ALPA, with increasing concern at the direction of the inquiry, issued a statement to the NTSB.

'It has been alleged during the investigation that the crew, through some unorthodox procedure, inadvertently extended the leading edge slats, recognised their mistake, and took action to retract them. This allegation further assumes that due to pre-existing damage to the number seven slat it did not retract, but went to the fully extended position. The crew members vehemently deny that this happened. TWA undertook a flight test to determine the effect of extending the leading edge slats at the same height and airspeed. According to the pilot of this flight, Captain George Andre, the aircraft experienced moderate buffet. This statement contradicts any possible extension of other than the number seven slat as the initial onset of this incident.'

At no time during the investigation, or at any other time, were the crew confronted with the evidence upon which the Board based its findings: the analysis of the flight date recorder, the results of the simulator exercises and the outcome of the flight tests. Captain Gibson and his colleagues had to await the publication of the official report before such details were revealed. It was to be June of the following year, 1981, and a further sixteen months of anxiety for the crew, before the NTSB's findings were released.

'The Safety Board determines that the probable cause of this accident was the isolation of the No. 7 leading slat in the fully or partially extended position after an extension of the Nos 2, 3, 6 and 7 leading edge slats and the subsequent retraction of the Nos 2, 3 and 6 slats, and the captain's untimely flight control inputs to counter the roll resulting from the slat asymmetry. Contributing to the cause was a pre-existing misalignment of the No. 7 slat which, when combined with the cruise condition airloads, precluded retraction of that slat. After eliminating all probable individual or combined mechanical failures or malfunctions which could lead to slat extension, the Safety Board determined that the extension of the slats was the result of the flight crew's manipulation of the flap/slat controls. Contributing to the captain's untimely use of the flight controls was distraction due probably to his efforts to rectify the source of the control problem.'

It was a tremendous blow to Flight 841's crew and one that was followed by vociferous denials from the three concerned, as well as hostile reaction from ALPA. But there was more. The report has been signed by three NTSB members, one of whom was Francis McAdams, the only pilot and aviation professional on the Board. McAdams was unhappy with some aspects of the case, especially the fact that neither he nor anyone else on the Board had seen or interviewed any of the accused, and he appended his own statement to the report.

'Although I voted to approve the Board's report which concluded that the extension of the leading edge slat was due to flight crew action, I do so reluctantly.

'The report as written, based on the available evidence . . . appears to support the Board's conclusion. However, I am troubled by the fact that the Board has categorically rejected the crew's sworn testimony without the crew having had the opportunity to be confronted with all of the evidence upon which the Board was basing its findings. At the time of the first deposition . . . no evidence was available to the crew or to the Board. Although the crew was deposed a second time, their testimony was limited to one issue, i.e. the physical location of the flight engineer at the time of the incident. I had recommended that since the Board was ordering a second deposition it be conducted *de novo* so that the crew would have been aware of all the evidence. The Board did not agree.

'Furthermore, I do not agree that a probable cause of this accident, as stated by the Board, was "the captain's untimely flight control inputs to counter the roll resulting from the slat asymmetry." In my opinion, the captain acted expeditiously and reasonably in attempting to correct for the severe right roll condition induced by the extended slat.'

The report, although firm in its conclusions, remained unconvincing, and the crew was left with a 'not proven' verdict. Under a cloud of uncertainty the three continued their duties with TWA, still proclaiming their innocence. The strain, however, proved too much for the second officer and Banks retired early to become a college lecturer.

As time went by the struggle to clear their names did not diminish and much information was gathered by ALPA in support of their cause.

At the beginning of the inquiry, suspicions of malpractice were initially aroused by what appeared to the investigators to be the deliberate erasure of the cockpit voice recorder (CVR) by Captain Gibson. The crew, the Board surmised, must have had something to hide. Erasure of the tape by Gibson, however, was clearly shown to be impossible. The CVR can be erased by the pilot only when the aircraft is safely on the ground with the parking brake set. Squat switches on the landing gears contact when the oleos compress with weight, closing circuits which confirm the machine is on the ground. The high-speed lowering of the landing gear during the upset resulted in damage which tore the squat switches and circuitry from the structure and the 727 landed with both main gear lights glowing red. The CVR erasure circuit was incomplete and, even if the captain had pressed the button on the flight deck, nothing would have happened. Gibson could *not* have deliberately erased the tape and the Board's initial suspicion that he did so because he had something to hide was unjustified. What is more, this was known and understood by the investigators early in the inquiry. Also after erasure, the tape recycled and a further nine minutes of conversation, including statements relating to the incident, were recorded. If it had been Gibson's intention to erase the tape, and it had been physically possible to do so, it is most unlikely that any evidence would have remained. So how was the tape erased? *Popular Mechanics* magazine interviewed a CVR technical expert and revealed that on a slow transfer of electrical power from the 727 engine to the auxiliary power unit (APU), which Banks accomplished on this occasion, erasure of the CVR is possible. This is the only way in which erasure of the CVR could have occurred. The timing of the tape recycling and the moment of slow power transfer also corresponded, as did the ending of the recording, nine minutes later, and the shut-down of the APU.

The question of whether the flight engineer had left the fight deck was also suitably resolved. Gibson and his colleagues denied that Banks had left the cockpit and it was established that shortly after top of climb the flight engineer had risen only momentarily from his position to place finished meal trays at the back of the flight deck for collection. Flight attendants, when interviewed, supported the flight crew's statements, and fourteen passengers, including four to five first class travellers who were in a good position to view movement from the flight deck, were also questioned. All swore that Second Officer Banks did not leave the flight deck near the time of the incident. The NTSB refused to consider this evidence and the statements were ignored. During the investigation, Captain Gibson kept hearing that someone had told the Board of Banks's absence, although the informer's identity could not be revealed, and it was realised only much later that it was Gibson's own comment which had led to the misunderstanding. The FAA representative who had spoken unofficially to the captain immediately after the accident, in spite of claiming that 'he was not wired for sound and wouldn't take notes', did write down in pencil the next day his recollections of the interview. It was the FAA representative's mistaken belief that Banks had left the flight deck, unwittingly reported to the NTSB, which had started the Board's 'what if the flight engineer had left the cockpit' scenario. With Banks on the flight deck the NTSB had no case.

The Board's version of the slat extension also did not stand up to scrutiny and when examined closely seemed absurd. The NTSB imagined the 727 levelling off at 39,000 ft (11,890 m), the power being set and Banks leaving the flight deck for a brief spell. Since all on board, except one lady passenger in economy class, testified to not having noticed his departure, his absence must have been short indeed. One of the pilots was then imagined jumping from his seat and in the darkness moving aside the crew's coats and swiftly finding and pulling the relevant circuit breaker. Very quickly, and in secret, the trailing edge flaps are set to 2°. Almost immediately Banks is imagined returning and noticing in the darkness, *beside* the crew's coats, that the circuit breaker is pulled and, *without saying anything*, pushing it back in again. According to the NTSB, half the leading edge slats then fully extend causing, by their own admission, moderate buffeting, and the three crew members are imagined sitting doing nothing for six seconds as they watch the leading edge flap lights sequencing. Then, with the leading edge slats fully extended, someone realises the error of their ways and is imagined selecting the flap lever up, thereby retracting all the lift devices except the number seven slat which inadvertently remains fully extended. This account, unfortunately, fails to appreciate the fact that if the pilots had just set the flaps they would have been highly aware of the circumstances and at the first sign of buffet one of them would have selected the flap lever to up long before full leading edge slat extension had occurred. It also fails to appreciate that modern airline crews do not operate in this manner and that in any usual situation, circumstances are discussed fully before action is taken. And, if the practice of selecting two degrees of flap had been widespread, as alleged, Banks, had he noticed the popped circuit breaker, would have known exactly what was going on and that would have been the last thing he would have touched. The NTSB scenario did not fit the facts, and even if the crew had deliberately set 2° of flap, an action not entertained here, the extension of number seven leading edge slat was highly unlikely to have happened in this manner.

There were also a number of other factors relating to the investigation which were disturbing. One hundred and eighteen simulator tests were undertaken by Boeing and the NTSB in order to resolve the situation but all tests were conducted according to the flight data recorder (FDR) and not one trial was performed in the manner described by the flight crew. The scientific officers analysing the accident data accepted only the scientific details as revealed by the FDR, even though it was known the equipment could be unreliable after being subjected to such a violent manoeuvre. The recorder indicated, for example, a rapid heading change of about five degrees to the right in less than a second at the start of the upset which was identified as a sharp roll. The flight crew denied that the wings rolled, stating that the upset was preceded by a fierce yaw to the right. The flight attendants also stated that they noticed no rocking of the wings and the passengers interviewed said the same. Not one simulator exercise was conducted with yaw as described by Gibson. All tests were performed in a 727-200 series simulator, whereas Captain Gibson's machine was a 727-100, a shorter version which was considered by pilots to be less stable at high altitude. Having taken the decision to fly at the unusually lofty level for a 727-100 of 39,000 ft (11,890 m), the crew were then highly unlikely to jeopardise further the aircraft's stability by deploying the trailing edge flaps. As Captain Gibson commented, 'It is very difficult to get three crew members to agree to take a 727-100 to 39,000 ft. I seriously doubt that the three most stupid pilots in the industry would ever consider experimenting with flaps at 39,000 ft in a 727-100.'

All NTSB trials were restricted to the suspected Board scenario and all fault analysis was confined to calculated problems which could have resulted in the extension of a slat. The more improbable cases were simply discarded. The US Air Line Pilots' Association (ALPA) demonstrated that 'a fracture of the actuator piston circumferentially through the lock key hole', no matter how unlikely, could have resulted in unscheduled extension of only the number seven slat. Although the NTSB and Boeing pointed out that the piston design strength negated the possibility of this failure and that it was claimed no record of such an occurrence existed, it was still possible, especially since the number seven slat mechanism had been damaged previous to the upset. The aileron hinge bolt that had also failed had a design strength which, likewise, negated the possibility of its shearing, and there was also no record of such a fracture existing, yet the bolt had failed. If one component which was deemed impossible to fail had done so, it was surely possible that another on the same aircraft could have suffered a similar fate.

The passengers were also asked if they had heard any unusual noises during the flight. The investigators hoped that some might have mentioned the unmistakable, high-pitched, shrill scream of the hydraulic motors operating which would have indicated extension of the trailing edge flaps. In the quietness of the cruise at 39,000 ft (11,890 m) the hydraulic motor noise would have been quite piercing, much louder than at the lower altitudes when flaps are normally operated after take-off or before landing. No-one, however, noticed any high-pitched scream, although many mentioned the light buffet and hearing the bangs of the compressor stall. This, alone, seemed conclusive proof that the crew did *not* operate the flaps.

The NTSB's accusations regarding untimely control inputs at the moment of the upset, disagreed by Francis McAdams, the only pilot on the Board, were also disputed. As early as 1975, Boeing had conducted a 727-100 flight test 'to investigate stability and control characteristics at high speed with one leading edge slat extended.' It is interesting to note that the test was ordered two years after the slat problem was assumed to have been solved by the issuing of the Airworthiness Directive in 1973, which required strengthening of the actuator piston rod ends. The aircraft took off with the number two slat bolted in the extended position and climbed in that condition to the required altitude. The test was abandoned at 33,400 ft and Mach 0.80 because of severe control difficulties resulting from heavy outboard aileron vibration. The test engineer at the time, D.L. Mahon, wrote on the flight log, as much in alarm as in jest, that the trial was curtailed 'due to several cases of extreme cowardice'. The aircraft, therefore, was never tested in this condition throughout its entire flight envelope. Captain Gibson's upset had occurred at 39,000 ft and Mach 0.81. Later it was demonstrated that the control forces at high altitude were not sufficiently effective to recover from a gross upset and that the denser air at lower levels was required for full control effectiveness. The performance of the simulator during the trials was based on wind tunnel testing which had not been verified by flight test at high altitude. In fact, the analytical data for flap extension at lower altitudes had had to be adjusted significantly to match flight test results. As a note, the severe roll to the left experienced by Gibson after selection of the remaining slats at about 13,000 ft (3,960 m) on the approach to Detroit could not be explained by Boeing. At slow speeds with the number seven slat missing and all other slats set, the aircraft should have rolled to the right.

During the Boeing 727-100 flight tests in October 1980, while checking performance in cruise with slats retracted and 2° of trailing edge flap extended, it was discovered, to everyone's surprise, that the configuration rapidly deteriorated performance. If the practice of selecting this configuration had been as widespread as alleged, crews would have been aware of the poorer performance; the corollary being, of course, that once the crews had discovered the deteriorated performance the practice would not have been widespread. Something, somewhere was wrong.

The entire investigation was conducted without the assistance of Captain Gibson and his colleagues and, after the initial testimony taken under the glare of publicity about one week after the incident, the flight crew were denied due process. At no time after the first deposition, during the whole two and a half years of the investigation, the longest ever, were the flight crew consulted, cross-examined or questioned by any investigator, except to answer 'no' to the question of whether or not Banks left the flight deck near the time of the incident. All of Gibson's phone calls to the NTSB were either refused, or never returned, and all notarised statements were ignored. The flight crew were also denied admission to the first hearing, which ended inconclusively, although the press and other interested parties were permitted to attend. Captain Gibson and First Officer Kennedy were admitted to the second hearing but they were not allowed to take part and they weren't even recognised by the investigators.

An appeal against the verdict was launched but it was to be eight years before it reached the Supreme Court, the highest authority in the land. Unfortunately, Gibson's case was not one of those selected for review and the situation was never resolved.

In 1983, a few of the passengers who had suffered minor injuries in the upset took TWA and Boeing to court in pursuit of compensation. Judgement was found in the plaintiffs' favour and sums of money were paid by both companies.

In spite of the result, however, Gibson and his crew, according to Donald Mark Chance Jnr, the lawyer representing TWA, came out of the proceedings well.

Captain Gibson and his colleagues attempted to take their case against the NTSB and Boeing to court but were denied due process of the law. A federal judge declared Gibson a 'public person' and, because of other legal difficulties at the time, action by the flight crew through the civil courts was refused. A documentary, entitled *The Plane that Fell from the Sky*, was made by Paul and Holly Fine for CBS's *60 Minutes* programme and was sympathetic to Captain Gibson's case, but was to no avail.

The crux of the matter in this entire case lay in the statement of the original hearing officer, Les Kampschror, that if the crew were not found in error then the airworthiness of the 727 would be cast in doubt. Since, in the minds of the investigators, no mechanical fault of the 727 appeared feasible, they left themselves little choice but to blame the flight crew. The majority of the NTSB were satisfied with the decision. The publicity surrounding the case and the verdict against the crew, it was reasoned, would eradicate the alleged widespread practice of tampering with the flaps in the cruise and further inadvertent slat extensions would be prevented.

On 28 August 1982, a 727-100 of International Air Service Company Ltd experienced an uncommanded number seven leading edge slat extension, after slat and flap retraction, while climbing through 4,000 ft (1,220 m) en route from Tulsa to St. Louis. Substantial aileron inputs were required to maintain control. Inspection revealed that the number seven slat actuator had malfunctioned and extended the slat.

On 17 November 1984, once again on the climb out with flaps and slats retracted, an unwanted extension of the number eight leading edge slat occurred to an American Airlines 727-200 whilst flying from Las Vegas to Los Angeles. A large lateral control input was required to control the aircraft.

In the cruise, on leaving 31,000 ft (9,450 m) for 29,000 ft (8,840 m), a partial and momentary extension of number two leading edge slat occurred to a United Airlines 727 on 10 September 1985. The aircraft pitched and yawed moderately. No cause of the unscheduled slat extension was discovered. Remarked the captain later, 'I know Boeing says it can't happen, but it did.'

On 8 January 1986, a TWA 727-100 outbound from St. Louis with flaps and slats retracted, experienced an unscheduled extension of the number seven leading edge slat whilst climbing through 6,000 ft (1,830 m). The aircraft rolled to the left with moderate buffeting and the captain decided to return to St. Louis. On the ground the number seven slat remained extended when the flaps and slats were selected up. Inspection revealed that the inboard number seven slat track extend stops were missing.

Whilst climbing out from Detroit on 22 February 1987, with flaps and slats retracted, an unscheduled extension of the number two leading edge slat occurred to a North West 727. The captain reduced the speed to 210 knots, selected two degrees of flap (which also extended slats three, six and seven, giving a symmetrical slat condition) and continued to destination, Cleveland.

Another incident occurred on 24 August 1987 to a TWA 727 en route to St. Louis. At a speed of 300 knots, and climbing through 10,000 ft (3,050 m), with the flaps and slats retracted, an unscheduled extension of the number seven slat began. The captain reduced speed and the number seven slat retracted of its own accord. Later, the number seven slat actuator was replaced.

In none of these incidents were the crew accused of tampering with the flaps. To this day, two decades later, the NTSB's verdict of guilty on Captain Gibson, First Officer Kennedy and Second Officer Banks, still stands. It is time the judgement was reviewed.

Note: 'Hoot' Gibson flew as captain on L1011s for three years after the upset, then transferred to the Boeing 747 as a first officer in 1982. In 1983, the litigation in which he became involved took its toll and in January 1984 Gibson retired early on medical grounds, with review after five years, and went back to his farm in Costa Rica. After only two years of farming 'Hoot' Gibson returned to flying duties with TWA in January 1986 as a 747 captain. Jess Kennedy remained in service with TWA and gained his command, but Gary Banks never returned to flying.

Chapter 9

Strange Encounter

The Islamic Festival of Ramadan falls in the ninth month of the lunar calendar and celebrates the first revelation of the Koran to Mohammed. The event begins with the new moon and ends with the old, and marks a month of daylight fasting for the Muslim faithful.

In 1982, 24 June fell on the second day of Ramadan and the night was moonless and dark. At just after 20:00 local time a British Airways Boeing 747, en route from the United Kingdom to New Zealand, took off from the Malaysian capital of Kuala Lumpur into the clear, black night. A quick five-hour hop to Perth in Western Australia lay ahead. For the passengers enduring the twenty-nine-hour journey from London to Auckland, Kuala Lumpur marked the half-way stage of the flight. The travellers from Europe destined for Perth were joined by a tour group of about thirty returning to that city who had been on holiday in Malaysia. Of the 247 total on board, therefore, about a hundred were due to disembark in Perth, the first Australian stop. The flight was then scheduled to continue via Melbourne to Auckland.

In 1982, British Airways won the 'Airline of the Year' award, being voted number one by regular business travellers, but the accolade had not been achieved without difficulty. Industrial and security problems at Heathrow had disrupted many departures and British Airways' Flight BA 009 from London to Auckland on 23–25 June was no exception. As the Boeing 747, named *City of Edinburgh*, registration *G-BDXH*, known phonetically as X-ray Hotel, lifted off from Kuala Lumpur's runway 15, the flight was running about one and a half hours late. The aircraft's weight at take-off was about 304 tonnes, which for the short journey was well inside the maximum structural take-off weight of 371 tonnes. The ninety tonnes of fuel carried was more than sufficient for the flight to Perth, which would consume about fifty-five tonnes, but with the Western Australian metropolis being one of the most isolated cities in the world more was needed in case of diversion. The designated second choice in case the destination airport was closed because of weather, or some other unforeseen circumstance, was the remote military base of Learmonth lying on the coast 600 nm north of Perth. With the high fuel load, however, BA 009 could, if necessary easily overfly Perth to Adelaide, a somewhat less isolated alternative. The briefing in Kuala Lumpur attended by the flight crew, Captain Eric Moody, Senior First Officer (SFO) Roger Greaves and Senior Engineer Officer (SEO) Barry Townley-Freeman, had given no indication of any adverse weather conditions en route, or in Perth, but the extra fuel carried was a sensible precaution. All expected a smooth, pleasant and routine flight.

On boarding in Kuala Lumpur, the passengers made themselves comfortable for the journey which lay ahead, but those travelling from London were becoming overtired and found it difficult to settle. Of the Australian tour group bound for Perth from Kuala Lumpur many were unhappy at being kept waiting for so long at the terminal before the late departure. Unconcerned with the problems at Heathrow the general feeling was that the Poms had let them down again. The big jet roared down the runway and after a long run because of the hot air became airborne at 20:09 local time (12:09 GMT). At 500 ft (150 m) in the darkness X-ray Hotel banked slightly left in the smooth air to turn towards Singapore and the route to Perth. Captain Moody established the aircraft on airway G79, flying towards the Johor Bahru VOR (very high frequency omni-directional radio-range beacon) which lay in the south of Malaysia at a distance of 150 nm. The climb continued steadily to the cruising altitude of 37,000 ft (11,280 m) and the aircraft was accelerated to the normal climb speed of 320 knots. The autopilot was now engaged and would remain so until approaching Perth. Passing 11,000 ft (3,350 m) the altimeters were adjusted to the standard pressure setting of 1013.2 millibars.

In the cabin drinks were served. At the request of the tired passengers the bar service was to be followed immediately by a light meal and then afterwards the lights were to be dimmed. In the quietness the weary travellers hoped to get some rest before arrival. As the journey progressed it soon became apparent to the passengers that they were being looked after by very able flight attendants, and they began to relax and make themselves at ease. Captain Moody's entire crew of sixteen, consisting of himself and the other two on the flight deck, plus thirteen in the cabin, led by Cabin Services Officer (CSO) Graham Skinner, had been together since they left London five days earlier. On this trip they had already been as far east as Jakarta, the Indonesian capital. The previous day the crew had then passengered out of Jakarta's International Airport, Halim, to position to Kuala Lumpur to pick up the BA 009 which they now operated down to Perth. The rapport enjoyed by the crew was reflected in the comments of their passengers, many of whom remarked that they seemed a 'happy band'. Little could any of these travellers have guessed at the time that before the flight was out they would be more than grateful for the professionalism of Captain Moody and his gregarious team.

As the climb continued the captain lifted the PA handset to say a few words to the passengers before they settled down for the night. He

A British Airways 747. *(British Airways)*

described the route the aircraft would follow as it flew over the Indonesian archipelago, explaining that from Singapore they would fly down the coast of Sumatra and across the Java Sea to overhead Jakarta, then from there they would proceed over the Indian Ocean to Carnarvon on the Australian coast, and on down to Perth. The remaining flight time of just under five hours gave them an estimated time of arrival of 01:30 local time in Perth, the time there being the same as in Kuala Lumpur. There was little or no cloud expected en route so flying conditions would be smooth, and the forecast for Perth was fine. The captain concluded by saying that he would leave them in peace now in the hope they might get some rest and he would talk to them again on the descent into Perth.

Approaching the Johor Bahru VOR beacon, X-ray Hotel levelled off at 37,000 ft (11,280 m) and accelerated to the cruise speed of Mach 0.85. The buffet speed of 265 knots, the lowest cruise speed for that flight level and weight, was bugged on the airspeed indicator (ASI), and the three-engine drift-down cruise level of 27,000 ft (8,230 m) was also noted. If an engine failed suddenly at 37,000 ft X-ray Hotel would be unable to maintain its altitude with reduced thrust and would have to descend to the denser atmosphere at 27,000 ft to maintain level flight. As the aircraft weight reduced with fuel consumption the three-engine drift-down level would, of course, rise and would eventually reach the actual cruise altitude if the aircraft remained level.

Over Johor Bahru, air traffic control (ATC) changed to Singapore, and BA 009 was cleared to proceed by airway B69 out of the Sinjon VOR beacon which lay just south of Singapore Airport. Airway B69 led all the way to Perth via Jakarta. Only the odd cumulonimbus (Cb) cloud could be seen flashing intermittently in the darkness below and the flight was proving to be as smooth and pleasant as expected. Captain Moody, SFO Greaves and SEO Townley-Freeman were being looked after by Stewardess Fiona Wright who had been assigned to duties in the first class upper deck area. The few first class passengers on board had all been accommodated in the main deck below and, with more than sufficient staff to cater for their needs, Stewardess Wright was free to attend to the flight crew. Normally they would have had to wait until after the passengers had been served but this evening they were able to eat right away. As the cockpit crew dined they remained strapped in their seats eating from trays on their laps. Duties were continued as usual with navigation and performance being monitored and communications being maintained en route.

Approaching Singkep radio beacon, situated on one of the outer Indonesian Islands, BA 009 was instructed by Singapore to call Jakarta on the HF long-range radio frequency of 6556 kHz. Meanwhile, in the cabin, the trays were quickly cleared away and within one hour of departure, as the aircraft passed abeam Palembang in southern Sumatra, most were quietly asleep in the dimmed light. A few minutes later X-ray Hotel entered the Jakarta upper control area and SFO Greaves established VHF short-range radio contact with Jakarta on 120.9 MHz. BA 009's progress was now monitored by radar, although position reports were still expected along the way. The crew were also monitoring their own radar sets, checking the small screens for any tell-tale signs of cloud activity lying in their path. A scanner in the nose was tilted down one degree to detect any weather ahead by reflecting transmitted signals from the large water droplets suspended in thunderclouds. The clouds would show up as small 'blips' on the pilots' radar displays. To avoid the Cb cloud it was simply a matter of engaging heading mode on the autopilot and steering round them. A faint outline of the northern coast of Java could be seen painted on the weather radar screen, but no large clouds were indicated on track.

An occasional flash of lightning appeared far to the east in the black night, and to the west lights on the coast of Sumatra were clearly visible. In the distance a low overcast obscured the brightness of the Indonesian capital. The Jakarta VOR was selected and the needles of the indicator pointed steadfastly ahead, indicating that the inertial navigation system (INS) was accurately flying the south-east airway track of 157° magnetic towards the beacon. At 126 nm north of Jakarta X-ray Hotel crossed position Bidak, an important reporting point which lay on the Indonesian air defence identification zone. The wind blew from the north-east at about thirty knots, giving a ground speed of 490 knots. The time was 13:17 GMT and it was estimated that X-ray Hotel would be overhead Jakarta in sixteen minutes' time.

'Jakarta, Speedbird 9', called Greaves on the radio.

'Go ahead.'

'Speedbird 9 was Bidak at 1317, level 370, Jakarta 1333.'

With the flight crew's meals finished the trays were removed and

Captain Eric Moody.

Stewardess Wright now took time to prepare her own fare. In the cockpit the relaxed progress of the flight left the crew at ease. In just under four hours they would be landing at Perth and after a short ride to the hotel some would be enjoying a few beers at a room party.

At 13:33 GMT BA 009 passed over the Jakarta VOR beacon, situated in a large bay to the north of the city, and from there the aircraft turned towards Halim VOR, lying only seventeen nm away. Its position six miles (9.6 kilometres) south-east of Jakarta marked the international airport from which the crew had departed northbound only the day before. At 13:35 GMT X-ray Hotel flew over Halim, still level at 37,000 ft, and Greaves once more radioed details of their progress to ATC.

'Speedbird 9, Halim at 35, level 370, Topar 52.'

The flight had been airborne now for almost one and a half hours and Captain Moody felt it was time to stretch his legs. On long-haul journeys flight crew are encouraged to take short breaks from the cockpit and all emergency drills and procedures are designed to be performed by only two people. SFO Greaves and SEO Barry Townley-Freeman were both very able and experienced airmen, with thirteen and eighteen years' flying experience respectively, so Captain Eric Moody, himself with 9,000 flying hours and seventeen years' experience, had no reservations about leaving his co-pilot in charge. Flight conditions were satisfactory and a check of the weather radar indicated a clear path ahead. On exiting the flight deck Moody found the crew toilet occupied and seeing no sign of Fiona Wright assumed it was she who was having a wash and brush-up before her own meal. He went below to the main deck but at the bottom of the stairs he bumped into the first class purser, Sara de Lane Lea, and stopped to talk to her for a while. He never did get to the toilet.

On the upper deck Stewardess Wright was at last able to begin her own meal, while on the flight deck all was calm. The quiet and routine progress of the aircraft reassured the two left in the cockpit that the flight was proceeding as planned, but it was an illusion of wellbeing that was soon to be shattered. In a few short moments the crew of BA 009 were to find themselves entangled in an emergency of dire proportions; caught up in an event unprecedented in the history of civil aviation.

At about 13:40 GMT Roger Greaves leaned forward in the right-hand seat to peer into the darkness beyond. His attention had been drawn by a strange visual effect on the windscreen and what appeared to be hazy conditions outside. A quick inspection of the radar screen showed no storm cells lying ahead and they were probably just catching some thin cirrus cloud or perhaps penetrating the tops of weak isolated cumulus clouds. He switched on a landing light to check the situation. At night it is sometimes difficult to tell if cloud penetration has occurred and such a procedure is an old pilot trick to illuminate the cloud if weather is encountered. A hazy effect seemed to be apparent and as a precaution the engine igniters and anti-icing systems were switched on. Combustion is normally self-sustaining once a jet engine is running, but heavy moisture can cause it to run down. In such circumstances igniters are employed which spark to maintain the engine lit without any adverse effect. Also, water droplets suspended in cloud can freeze on impact with engine sensors and nacelles, and the anti-icing systems prevent disruption of the airflow through the engine. Flashing the landing light occasionally would indicate when the aircraft was in the clear and both kept a weather eye open for any change in the circumstances. It was not long before their vigilance was rewarded.

'Hey Barry, look at this.'

SEO Townley-Freeman had also noticed a glow and was already looking at the sight. Streaks of mini forked lightning flashed across the windscreen. The electrical effect is known as St Elmo's fire, which both men had witnessed before in their careers, but it is normally associated with electrical discharges in storm clouds. The flight remained smooth, nothing could be seen ahead in the form of flashing thunder clouds and no adverse weather returns appeared on the radar. All visual indications, however, were that BA 009 was on the edge of a storm cloud of intense electrical activity. Giant Cb cloud in the region can stretch up to 50–60,000 ft (15–18,000 m) and can give an aircraft quite a shaking, but it was most unusual for such thunderstorms not to appear on radar screens. If the flight was about to penetrate a storm cell it was as well to be prepared and SFO Greaves switched on the 'fasten seat belts' sign. Both flight crew already had lap and crutch straps fastened, but each now attached his shoulder harness.

The St Elmo's fire continued to develop and the two left on the flight deck had a front-row view of a spectacular kaleidoscope of streaking light. The co-pilot called Stewardess Wright on the interphone.

'Fiona, come and see this. It really is quite a sight.'

'I've been flying too long to fall for anything like that, Roger. I'll be up in a minute.'

'No, no,' he insisted, 'come now. You won't believe your eyes.'

As BA 009 approached the south coast of Java, Fiona entered the flight deck and walked forward to stand behind the captain's empty seat. She stooped to look at the windscreen and was immediately taken by the beauty of the sight. The whole aircraft seemed to be enveloped in a glow and the scene ahead was like skiing at speed through a snowstorm of bright silver sparks. Amid the recognisable flashes of St Elmo's fire, streams of light like tracer bullets streaked up the windscreen. In the half-light of the darkened cockpit their faces almost shone in the irridescent light. But there was more. A strange ionised odour, like that associated with electrical sparking, could also be detected and a thin veil of mist with a bluish hue began to shroud the flight deck. Something very odd was happening to BA Flight 009.

'I don't like the look of this', remarked Greaves. Quickly he turned to Stewardess Wright. 'I think you'd better get the skipper back up here.'

Fiona Wright made her way down the spiral staircase with some haste and from half way down called out to Captain Moody who was still standing chatting by the galley at the foot. He had been joined by CSO Skinner.

'Captain, you're wanted on the flight deck.'

He didn't need to hear her request twice to detect the urgency in her voice. Stewardess Wright turned and hurried back to the upper deck while Moody immediately dashed for the spiral staircase, mounting the steps two at a time. CSO Skinner followed behind. As the captain's head reached the floor level of the cabin above, however, he almost stopped in his tracks. In the dim light he could see from below the seats what appeared to be puffs of smoke emanating from the low-level air conditioning ducts. It reminded him of the clouds of water vapour that billow from the air conditioning vents in hot and humid climates. He, too, could detect the acrid electrical odour. 'Smells like the London underground', he thought to himself. It could also be an indication of an electrical fire. Any fire in flight is one of the greatest hazards to be encountered and he was now not at all surprised that he had been summoned back to his post.

He rushed into the flight deck and rapidly scanned the cockpit. The engine anti-icing and igniters were checked on and the fuel panel configuration was checked as satisfactory. Although the problem seemed to lie outside nothing could be seen on the radar.

'It's quite incredible', said Greaves.

Captain Moody stared wide-eyed at the intense and spectacular display of dancing lights. For a moment the St Elmo's fire, tracer and glowing luminescence distracted his attention from the smoke enveloping the flight deck. The two cabin crew, Stewardess Wright and CSO Skinner, stood behind, almost unable to believe the beauty of the sight.

'And look at the engines', said Roger Greaves.

The captain turned to the left and could see what appeared to be the engines lit from within by an intense white hot light, like the illumination from a magnesium flare. Shafts of light extended forward from all four engines. The powerful brightness shining through the fan blades produced a stroboscopic effect which made the fans appear to turn backwards in the engine nacelles. Barry Townley-Freeman also rose briefly from the flight engineer's seat to peek from the side window behind Greaves while the flight attendants took turns to view from the captain's side.

In spite of the impressiveness of the display, however, there was mounting concern for the 'smoke' on the flight deck which was becoming more dense. If it became any worse they would have to don oxygen masks. CSO Skinner didn't need to be told something was amiss, and he quickly left the cockpit to warn the crew to prepare the cabin in case of problems. He was met by other flight attendants reporting evidence of smoke throughout the aircraft and the instruction was given to switch off galley electrical power and to stow equipment safely.

To SEO Townley-Freeman, the thickening acrid smoke from the air conditioning system suggested an electrical problem, and he carefully scanned his large instrument panel for any signs of trouble. The time was now 13:45 GMT and X-ray Hotel was crossing the Java coast and heading out over the Indian Ocean on Airway B69. As Townley-Freeman continued with his scrutiny, he carefully opened the check list at the drill for removal of air conditioning smoke. The procedure initially called for the donning of oxygen masks and goggles by the crew and for the switching off of gasper and recirculation fans. On the captain's command he would be ready for action. Suddenly the flicker of a light from an unexpected source caught his eye. The intermittent flashing was from a pneumatic valve light which was designed to illuminate when the valve was closed. The valve supplied compressed air direct from the engine compressors to feed air-driven equipment and air conditioning and pressurisation systems. It was designed to close automatically when an engine ran down to prevent reverse flow, and its movement towards the shut position indicated more of an engine problem than trouble with the air conditioning system. Quickly Townley-Freeman glanced forward to check the main engine instruments situated in the centre of the pilot's panel. Number four engine on the outer right side was running down! Something had disturbed the flow of air through the compressors and the

The three flight crew: SFO Roger Greaves (left), Captain Eric Moody (centre) and SEO Barry Townley-Freeman (right) stand before *G-BDXH*'s number one engine.

engine had surged, i.e. had suffered a giant hiccup, and the power was dropping. By contrast the exhaust gas temperature (EGT) was soaring and the indicator pointer was racing off the clock. The amber EGT warning light illuminated.

'Engine failure,' shouted Townley-Freeman, 'number four.'

The Pilots also felt a slight swing of the nose as power dropped on one side and rudder was applied to hold the aircraft straight.

'Engine fire drill number four engine', responded Captain Moody.

Greaves and Townley-Freeman carried out the fire drill from memory, closing the number four thrust lever, shutting off the fuel supply and pulling the fire handle to isolate services from the failed engine. As they completed the drill and confirmed their actions from the check list, the captain monitored the flying, holding on a little left rudder with his foot to balance the aircraft. The entire procedure had taken about thirty seconds. Some fairly quick thinking was now required, for on three engines X-ray Hotel was unable to maintain 37,000 ft (11,280 m). At about 280 tonnes they would need to descend down to 29,000 ft (8,840 m), the appropriate level for their weight and flight direction. Permission would be required from Jakarta and would have to be obtained promptly. It would also be unwise to commence a journey over the ocean on three engines, for if a second engine failed in mid-crossing BA 009 would be in a serious situation. Jakarta lay about ninety nm to the north and, with flight easily sustainable at the lower level on three engines, the sensible approach would be to turn back and land at Halim to have the engine checked. A rerouteing, however, would also require clearance over the radio. Before the crew could plan or even think procedures, events took a turn for the worse. Number two engine surged and began to run down. Once again the power drop was accompanied by a rising exhaust gas temperature.

'Number two's gone', called the flight engineer.

Almost immediately Townley-Freeman saw to his horror that numbers one and three engines had also surged and were running down.

'They've all gone!' he shouted with disbelief.

If there was alarm on the flight deck at the situation, in the cabin the passengers, unaware of the problem, were becoming increasingly distressed. Although the main cabin lights were dimmed, the acrid 'smoke' in their midst was quite obvious and the thumping of the surging engines could be heard and felt through the structure. The glow from the engines could be clearly seen in the black night and huge sheets of flame could be seen shooting from the jet effluxes illuminating the interior of the cabin. To those by the windows it seemed as if the whole aircraft was on fire. Suddenly, for a reason they could not comprehend, those at the rear of the aircraft noticed a drop in the noise level and they could sense the cabin being engulfed in an eerie silence.

In the cockpit the initial reaction of the flight crew was one of total disbelief. Four engines don't just stop; what have we done wrong, was their first thought. Engine intakes can freeze up, causing run-down, but the engine anti-icing had been checked on. So, too, had the igniters to sustain combustion. Fuel mismanagement can also cause problems: it had been known on the rare occasion for all four engines to be run from one fuel tank for balancing purposes and for inattention to have caused the tank to run dry, stopping the engines. Moody and Townley-Freeman had both confirmed the fuel configuration as satisfactory. The engine indicator pointers were all over the place, causing confusion, but there was no doubt that, for whatever reason, all engines had failed.

Crew training, of course, covers every contingency, no matter how remote, and total engine failure was no exception. Only a few months earlier Captain Moody had successfully completed his biannual simulator check and had practised this very procedure. The simulator cockpit had quietened, the lights had dimmed to emergency level, the autopilot had disconnected and many instrument failure flags, including most of those on the co-pilot's side, had appeared. On X-ray Hotel's flight deck, however, the situation was quite different. The rush of air masked any reduction in engine noise, the cockpit lighting stayed on, the autopilot remained engaged and the instrumentation, apart from the engine indicators, displayed as normal. Whatever was happening to X-ray Hotel was a complete mystery. The strange storm producing the intense electrical activity also seemed to be affecting the power, and some unknown force had choked the engines into silence.

Immediately Captain Moody instructed Greaves to send out a mayday, and Townley-Freeman commenced the 'loss of all generators' drill. The procedure assumed, of course, that primary services had been lost: pneumatics, including pressurisation and air conditioning, normally supplied by bleeds from the engine compressors, and electrics, normally supplied by the engine-driven generators. With total engine failure, however, no attempt to restart could begin without standby electrical power from the batteries to supply essential engine controls: fuel valves, start levers, standby ignition and some engine indicators. The 'loss of all generator' drill, therefore, began by checking that the battery switch was on and that the standby electrical busbars were powered. DC power from the battery was changed by an inverter into AC to supply the standby AC busbar. The combination of DC and AC buses also fed the captain's flight instruments, number one VHF radio, the interphone and the PA system. Having checked these items Townley-Freeman then opened the fuel cross-feed valves. In a matter of seconds he had reached the point at which the drill split: loss of electrical power caused by a massive short which had tripped all the generators, or the loss of all generators caused by total engine failure. In this case the course of action was obvious, and he prepared for in-flight starting of the engines. The air was too thin at cruising altitude for an engine start to be recommended, but under the circumstances there was little point in waiting and it seemed worth a try.

'Mayday, mayday, mayday', shouted Greaves over the radio. He called on the Jakarta control frequency of 120.9 MHz, having set the emergency code of A7700 on the transponder. If there was no reply he would transmit on the emergency frequency of 121.5 MHz. 'Speedbird 9, our position is 100 miles south of Halim. We have lost all four engines. We're descending and we're out of level 370.'

Silence, and then a few moments later: 'Jakarta, Speedbird 9, have you got a problem?'

The electrical static produced by the storm was distorting transmissions and making communications difficult.

187

'Jakarta, Speedbird 9,' repeated Greaves carefully, 'we have lost all four engines. We are descending and we are now out of level 360.'

Another short pause followed, then Jakarta replied. 'Speedbird 9, understand you have lost number four engine?'

Greaves turned in frustration to his Captain. 'The fuckwit doesn't understand.'

'Negative, Jakarta, Speedbird 9,' spoke Greaves slowly and distinctly, trying to control his frustration, 'we have lost all four engines, repeat all four engines. Now descending through level 350.'

The Jakarta controller was still unable to understand but fortunately a Garuda Indonesian Airways flight on the frequency interrupted the exchange.

'Jakarta, Garuda 875, Speedbird 9 has lost all four engines, he's lost *all* his engines, and he's out of level 370.'

At last Jakarta got the message and understood the full force of the problem. Greaves requested radar assistance for return to Jakarta but the conditions prevented the controller from identifying X-ray Hotel. In the end the first officer simply told the controller they were turning back for a landing at Halim Airport. As Greaves battled with the radio, Townley-Freeman fought to bring the engines back to life. The number four engine had been fully shut down so the fire handle was reset and the thrust lever was placed in line with the others. Although it was not advisable to relight an engine after a non-recoverable stall unless a greater emergency occurred, this occasion surely qualified. It seemed common sense to Captain Moody and the crew to attempt to restart any one out of four engines instead of any one out of three. For the flight crew, however, the mental action of reading the check lists and performing the drills in such a desperate situation was, in Barry Townley-Freeman's own words, like trying to think through treacle. In such circumstances it is the man who can use his common sense who will win the day. The reinstating of number four engine may have seemed like a simple act but it was to prove significant at a later stage.

To begin the start sequence the start levers were selected to cut-off to reschedule the fuel control units. The standby ignition was switched on to power the igniters and the start levers were selected back to idle. The crew looked in hope to see the instrument readings rising indicating light-up. Nothing. Not even a hint that at least one engine was going to start.

In the meantime, the captain had exchanged height for speed and had commenced descent to stop the speed dropping further. Using the autopilot he had set 500 ft (150 m) per minute descent and the speed had settled on the airspeed indicator at 270 knots, which was close to the speed for minimum drag for the present all-up weight. He now turned the autopilot knob to bank left, clear of the airway, and although X-ray Hotel was descending over the sea he set the safety height for the return to

Jakarta over Java of 11,500 ft (3,500 m) in the altitude select window. It is normal in such reports of aviation emergencies to read of aircraft plummeting to the ground, but in this case it was far from true. Normal descent procedures approaching destination include closure of the throttles to idle power at top of descent, and the aircraft literally glides to the lower levels before power is reapplied for the approach to land. The aircraft's sleek aerodynamic lines reduce air resistances to a minimum and even slowing down without power at level flight takes time.

In X-ray Hotel's situation the lack of any residual thrust from idling engines and the drag from the big engine fans windmilling in the airstream would increase the rate of descent, but it would only be marginal. The windmilling engines, however, were a distinct advantage in this emergency, for even at speeds as low as 160 knots they could still turn the hydraulic pumps sufficiently to pressurise the system for operation of the aircraft's flying controls. Also, unknown to the crew at the time, in spite of complete engine failure number three generator had stayed on line, supplying electrical power. It was being driven by the windmilling engine and was the reason the autopilot had remained engaged and the cockpit lighting and instrumentation displays had stayed normal. The generator would not remain on line for long, however, but when it tripped electrical power would be supplied by the batteries for thirty minutes.

Meanwhile the captain could use the autopilot to control the aircraft in the descent and leave himself freer to concentrate on the problems in hand. The cabin was normally pressurised by the engine compressors to a pressure equivalent to 6,000 ft (1,830 m), but with power loss, even though the outflow valves had automatically run closed, the cabin pressure was beginning to drop.

In spite of total engine failure, therefore, the aircraft was under control. Hydraulic pressure was sufficient to operate the flying controls, electrical power, although fluctuating, was still available from number three generator, and the cabin pressure, although reducing, was still sufficient for normal breathing. X-ray Hotel was descending at around 2,000 ft (610 m) per minute and was now out of 35,000 ft (10,670 m). In its clean condition, i.e. no flaps or gear lowered, it could remain airborne without power for about a further twenty minutes, covering a ground distance of approximately 140 nm. Although the aircraft was in dire straits the problems were compounding gradually and the crew were able to gather their senses. The initial alarm had subsided and with the adrenalin flowing the three worked well together as a team. Even the early failures to start the engines did not leave the flight crew in complete despair. Engine start required conditions to be within certain parameters: indicated airspeed 250–270 knots, altitude below 28,000 ft (8,530 m), and windmilling engine compressor rotations within tolerance. Once they had descended to lower altitudes they were sure the engines would fire up again. After all, the same had happened in the simulator check; once below 28,000 ft the engines had relit. And X-ray Hotel was also a Rolls-Royce aircraft: American-built but fitted with good reliable British Rolls-Royce engines. Of course they would start.

Turning the aircraft back to Jakarta brought BA 009 closer to a suitable airport where the captain could land once the engines had been restored, but attempting a dead-stick landing, i.e. touching down without power, could not be considered. Judging such a landing in the darkness would be almost impossible and the chances of success negligible. There was also a tall mountain range lying across Java which resulted in the safety height being as high as 11,500 ft. X-ray Hotel could not clear the high ground without power. Many of the mountains were volcanoes and some of them were very active. In the Sunda Strait which separates the Indonesian islands of Java and Sumatra was once situated the then volcanic island of Krakatoa which erupted on 27 August 1883. The explosion was equivalent to a force of 50,000 H-bombs and blew the island completely off the map. The bang went on record as the loudest produced on earth and was heard half-way round the world. Almost one hundred years later, another mountain of the chain, Mount Galunggung, situated 100 miles (160 kilometres) south-east of Jakarta on the Java coast and one of sixty-seven currently active volcanoes in Indonesia, had recently been belching fumes into the atmosphere. It was considered the most dangerous and a major eruption was expected.

If the crew were unable to start the engines they would be forced to ditch in the black sea, but that was some way off. There was still about eighteen minutes to save the situation and all their efforts and thoughts could be concentrated on the single aim of getting the engines going again. They had no opportunity to think beyond their immediate tasks and they had not reached the point of assuming the worst. At that stage they were too busy to be afraid, but they were not to realise then that shortly the situation would turn much worse. Later they would know fear.

Once more the three scanned the flight deck for clues as to their predicament: engine anti-icing on; standby igniters on when required; fuel panel checked, all main boost pumps on; circuit breakers checked, all confirmed set. The thrust levers were simply left in the positions they were in at the time of failure. Again the start levers were selected to cut-off, fuel shut-off valves were checked open, standby ignition was selected and the start levers were set to idle. Still no response.

By now the crew had exhausted all their options and it was simply a matter of selecting the start levers to cut-off, switching on standby ignition, and reselecting the start levers to idle. Standard procedures in an emergency, requiring the captain to fly the aircraft and the other two to perform the drills, were abandoned in the circumstances. With the autopilot still engaged and the radio work completed for the moment,

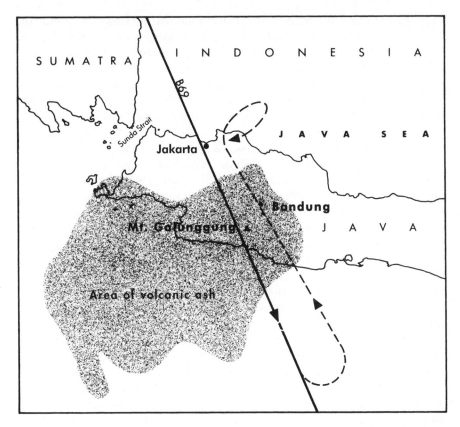

A map showing BA 009's routeing on B69. Also shown is the position of Mount Galunggung and the area of volcanic ash.

both pilots found themselves free to lend a hand and all three took turns at trying to start the engines. If persistence alone could have achieved results the four engines should have roared into life, but it was not to be. They remained ominously silent. In the cabin the initial alarm had subsided, but there was mounting apprehension at their plight. The captain had been too busy to make an announcement and the passengers were ignorant of the circumstances. Those at the rear of the cabin by the windows could witness, without realising what they saw, the crew's attempts to get the engines going. On each occasion when fuel was introduced and the engine failed to light-up, the kerosine ignited in a huge flame which shot from the jet efflux. The process shook and vibrated the aircraft.

The number three generator was still being driven by the slowly rotating, windmilling engine via a constant-speed drive which was unable to perform its task properly. The resulting fluctuating frequency made the cabin lights cycle on and off, brighter one moment then dim the next. At the back of the cabin the eerie silence also persisted and they could hear each other breathe. Under the circumstances it would not have been surprising if the passengers had panicked, but they remained remarkably calm. The cabin crew did a wonderful job in reassuring the travellers and moved about the cabin comforting those in greatest distress.

At this stage BA 009 was only a few minutes into the emergency and the crew seemed to have exhausted all their options except the hope of starting the engines at a lower altitude. Amazingly, most of the captain's and first officer's flight instruments remained powered and displayed normal indications and the cockpit lighting remained illuminated. Only occasionally did some engine instrument failure flags appear. The engine indicators were a mix of Smiths and General Electrics, and the former gauges froze where they stood with power loss while the latter gauges ran to zero. It made it almost impossible to properly monitor the engine start sequence and difficult to know whether or not light-up was occurring. To add to the difficulties, the electrical storm still danced magnificently before their eyes and their headsets crackled loudly from static interference.

X-ray Hotel was now descending through 30,000 ft (9,144 m) out over the ocean at a rate of 2,000 ft per minute, in a gentle turn back towards Jakarta. If the engines failed to relight Captain Moody would simply turn back southbound away from the mountains and remain over the flat sea. Townley-Freeman continued with the start attempts and with the approach of 28,000 ft (8,530 m) all were confident of a recovery. Still nothing happened.

With time, the bluish mist in the cabin began to disperse as the inert compressors of the failed engines no longer pumped contaminated air, or any air at all for that matter, into the aircraft. Oxygen masks were not needed as a protection against the 'smoke', but they would shortly be required for another reason. As X-ray Hotel descended through 28,000 ft (8,535 m), an intermittent warning horn sounded in the cockpit: *beep, beep, beep*. Although the aircraft was descending the cabin was, in a sense, ascending, as the pressurised air within leaked to the atmosphere. On 'climbing' through 10,000 ft (3,050 m) the cabin pressurisation warning horn sounded and the flight crew were obliged to use oxygen. It was a moment that those in the cockpit did not relish for the masks were uncomfortable to wear, but what happened next took them all by surprise.

To begin with the flight engineer's mask stowage had been incorrectly installed and it was impossible to reach while seated. At a time when seconds were precious he had to undo his seat belt and stand up before stretching backwards to reach his mask. As the pilots pulled their oxygen mask harnesses over their heads, the supply hose of Greaves's mask came away in his hand and the connecting pieces fell apart in his lap. He was now without oxygen in the 'climbing' cabin and as the altitude increased he was in danger, if the situation continued, of suffering from anoxia. The effect induces a feeling of wellbeing in the victim although performance is akin to being drunk. Unconsciousness eventually results.

Standard procedure for the failure of pressurisation in the cruise is for a steep descent to be initiated immediately. The throttles are closed, speed brakes are deployed, the gear is lowered and the nose is pushed down to attain maximum speed. Descent rates of 10,000 to 12,000 ft (3–3,600 m) per minute can be achieved and the aircraft can quickly be brought to safety at a lower level. In X-ray Hotel's case it was pointless to throw away precious height if, for the time being, the crew could breathe normally, but with the threat of incapacitation to his co-pilot, the captain had little choice but to get down quickly. Captain Moody disconnected the autopilot to hand-fly the aircraft and lowered the nose.

It was an agonising process for Moody to throw away height, the one commodity which could both buy them time and save everyone's life. The landing gear could be lowered by using the alternative system, which simply released uplatches to allow the wheels to drop by gravity. Once lowered in such a manner, however, it could not be raised. If a ditching became necessary, touching down on the sea with the landing gear extended could place the aircraft in danger. Moody compromised on procedures by pulling only the speed brake lever, but even so the aircraft dropped at an alarming 6,000 ft (1,830 m) per minute. The passengers could feel a rumble through the cabin with the speed brakes deployed. As height tumbled from the altimeters, Greaves fought desperately to refit the hose attachment. With nimble fingers and cool nerves he managed to fit the pieces together and to place the mask on his face. By now the aircraft had descended to about 20,000 ft (6,100 m) and Moody retracted the speed brakes to resume the glide.

'I've got 320 knots,' said Greaves; 'we are going too fast.'

A speed of 250–270 knots was required for engine start but his airspeed indicator displayed a higher speed. The captain quickly glanced across.

'That's strange, I've got 270 knots.'

The discrepancy of fifty knots was a mystery, but could be the cause of the engines failing to relight at the lower altitude. Captain Moody varied the speed using the co-pilot's airspeed indicator and then his own while the flight engineer repeated the start sequence, but to no avail. In the cabin, to add to the passengers' discomfort, the aircraft now seemed inexplicably to climb and descend at varying rates and they could feel the heaving in the pits of their stomachs. There was still no time for the captain to explain the predicament, even if he knew what was happening, so in more ways than one the passengers were left in the dark.

The cabin continued to 'climb' in spite of the aircraft descending and, as X-ray Hotel passed 18,000 ft (5,485 m), the cabin height rose through 14,000 ft (4,270 m). Passenger oxygen masks dropped from their over-head stowages. The appearance of the masks added to the distress of the travellers who had just felt what seemed to be the aircraft plummeting out of control. The cabin crew donned portable oxygen equipment and performed a marvellous job in calming the more disturbed passengers and helping to fit masks as they walked round the cabin. Unbelievable as it may seem, some travellers were still asleep. The emergency announce-ment explaining how to use the masks had failed to function and CSO Skinner found he was unable to operate the PA system from the cabin stations. He pulled a megaphone from one of the overhead lockers and attracted the attention of the passengers to give them instructions.

'Can you hear me, mother?' he called, imitating a northern English accent. It brought a moment of light relief that was greatly needed.

On the flight deck attempts to start the engines continued without success as Moody hand-flew the aircraft. The fifty knot discrepancy in the airspeed indicators was still evident and once again he varied the speed. X-ray Hotel was now running out of height although mysteriously descent was not as fast as expected. The outside air temperature was much higher than normal and the hot rising air was also producing an uplift to help keep BA 009 airborne, not unlike that experienced by a glider in an updraught. X-ray Hotel was still losing the race against time, however, and the flight crew were under a lot of pressure. The dim cockpit lighting, the wearing of uncomfortable oxygen masks, the flashing electrical discharges, and the loud headphone static did not make their tasks any easier. The electrical activity in the atmosphere also seemed to be playing some very strange tricks on the aircraft's electrical systems. The inertial navigation system displayed random digits and patterns, with nothing making sense, the VOR needles spun round in circles and the distance-measuring equipment (DME) was blank. The dark night made visual flying impossible and the captain had to fly solely on instruments.

It was now that nagging doubts began to invade their minds and thoughts that the engines might not start entered their heads. Townley-Freeman surmised that water contamination in the fuel tanks could be causing the problem and if that was so there was no way the engines were going to light up again. There would be no choice but to ditch. Captain Moody had already decided that he would fly towards Jakarta until approaching 12,000 ft (3,660 m), just above the Java safety height, and then he would turn southbound to land on the sea. It was not a prospect any of them relished. The pilots understood the principles of ditching: landing along the line of the primary or predominant swell and upwind and into the secondary swell, or downwind and down the secondary swell. That was the theory, of course, but landing on a shark-infested ocean in a black night with a heavy sea running was a different matter.

The captain had watched flying boats as a boy and knew that they didn't operate at night because of the difficulty in judging height in the darkness. On X-ray Hotel, if only battery power was available for ditching, there would be no radio altimeter for precise height indication and there would be no landing lights available. The landing gear would remain retracted and the underside of the aircraft would present a smooth surface to the sea on touchdown, but it would not be possible to lower flaps and the ditching would be fast. The aircraft stall speed at sea level was 179 knots, so to achieve a safe margin a speed of 190–200 knots would be required. The engines would almost certainly break off on impact and the wings and structure might be damaged. Once stopped the aircraft would probably remain afloat for a reasonable time, but launching the life rafts in the heavy swell would be extremely difficult. There would also be a very real danger of sharks. X-ray Hotel was now descending through 16,000 ft (4,880 m) and if the engines didn't start soon they were going to be in trouble. Strangely, the flight crew as yet felt no great fear, but there was certainly growing apprehension.

In the cockpit the captain now felt they had caught up with events but there seemed little more they could do. Townley-Freeman had not rested in his attempts to start the engines and he continued to try. He would still be having a go when they touched the sea, if it came to that. Greaves repeated requests for radar assistance from Jakarta, but the electrical storm was making contact difficult. There was still no range or bearing information from Halim VOR, but one of the three inertial navigation sets suddenly made some sense and true track and distance were observed and transmitted to Jakarta.

'We're out of level 150 now, descending', radioed Greaves, giving the position information at the same time.

'Roger, Speedbird 9, if possible maintain not lower than 12,000.'

It was a useful reminder of the safety height.

In these dire circumstances the captain felt the need to speak to CSO Skinner and to ensure that he understood the predicament, but Moody was unable to contact him on the interphone. The captain also felt that, in spite of the situation, he had some spare thinking capacity and now seemed an appropriate moment to talk to the passengers. He could also request Skinner's presence at the same time. Moody took a deep breath and spoke into the oxygen mask: there was little he could do but tell the truth.

'Ladies and gentlemen, this is your captain speaking. We have a small problem. All four engines have stopped. We are doing our damndest to get them going again. I trust you are not in too much distress.' There was a short pause in which the message took effect, then he added, 'Would the CSO come to the flight deck immediately, please.'

The last statement was not only a summons to Skinner, but also a signal to the cabin crew in an emergency to stow their equipment and to move to their respective stations. No one was now left in any doubt as to their predicament. As CSO Skinner made his way towards the flight deck, Captain Moody's mind flashed to a scene from a training film depicting a moment of great emergency. He could recall the amateurish screen captain delivering his instructions to his senior flight attendant in stilted but impeccable English.

'It seems it's not our day. I'm afraid we're going to have to ditch.'

It was not Moody's way. As Skinner entered the flight deck carrying his portable oxygen set the captain turned briefly to glance at his CSO.

'Got the picture?' he shouted with difficulty through his mask.

Skinner only heard a muffled sound, but didn't press the matter. He could guess from the activity on the flight deck that a forced landing, probably on the sea, was imminent. He gave a quick thumbs up to his captain and when no further direction was forthcoming he returned to the cabin.

The flight crew now began to feel more than concern for the height being lost. As the aircraft descended through 14,000 ft (4,270 m), about one minute remained before Captain Moody planned to turn southbound again and back out to sea. Once heading away from Java, approximately five minutes – no more – would be left for some further endeavour at starting the engines, then the flight would be over. The time was now 13:57 GMT, and X-ray Hotel had been without engine power for over twelve minutes. The crew had lost count of the number of failed attempts and it was unlikely at this stage that their efforts would bear fruit. It did not stop them trying, but a ditching seemed inevitable. Now they knew fear.

At that moment X-ray Hotel appeared to break from the hazy cloud of the electrical storm, and the aircraft flew into clear air. The dancing lights vanished from the windscreen and the glow from the engines dimmed. Suddenly Townley-Freeman let out a cry.

'Number four's started!'

The three watched with bated breath as the engine gauges rose steadily and the power settled. Gingerly Moody advanced the thrust lever and the engine ran successfully at normal power. It was only one relit, but they were on the road to recovery. In the cabin the roar of the engine was music to the passengers' ears. If they now turned out to sea and descended to a lower altitude, quickly dumping fuel at the same time to lose weight, they might just be able to remain airborne. If they couldn't hold height the power would at least slow the rate of descent. More importantly, the vital services of electrics, hydraulics and pneumatics could now be supplied from this one engine, and, if necessary, compressed air could be tapped from the engine to help turn the others and give a better chance of starting them. Number four engine had been shut down early in the emergency and it had been the least affected by whatever had caused the damage. It now seemed to have saved the day. The crew could also now remove their rather uncomfortable oxygen masks. X-ray Hotel's rate of descent eased to 300 ft (90 m) per minute, but the aircraft continued inexorably downwards. Approaching 12,500 ft (3,810 m) there was little margin to clear the mountains for an emergency landing at Jakarta. As the seconds ticked by the initial relief abated, for all efforts to get the others going proved to be in vain. One minute passed in what seemed like an eternity, and still the moments slipped away. A further twenty seconds went by without success, then Townley-Freeman called out again.

'Number three's lighting up.'

Soon number three was restored to life, and with two engines running at sufficient power to arrest the rate of descent, the aircraft was held level at 12,000 ft (3,660 m). X-ray Hotel could now remain safely at altitude if only the engines, which were probably damaged, could sustain power. They could then cross the mountains and execute a two-engine landing at Jakarta. Almost immediately numbers one and two engines relit in sequence and, from a situation of great danger only a few moments earlier, they now found themselves with all four engines running normally.

'Jakarta, Speedbird 9,' called a relieved Greaves, 'we're back in business, all four engines running, back at 12,000.'

Unfortunately the controllers were still unable to establish radar contact because of X-ray Hotel's proximity to the mountains, and a request was made for climb to 15,000 ft (4,570 m). As the aircraft levelled off at the new height, Greaves relayed their position, giving bearing and distance from Halim VOR, and radar contact was confirmed. The airspeed indicators had settled down and were reading in agreement but 'A' autopilot would not re-engage. 'B' was selected instead. They now seemed home and dry with their troubles over, but there was more to come. In the meantime the captain took advantage of this quiet moment to speak to the passengers.

'Ladies and gentlemen, we seem to have overcome our problems and have managed to start all the engines. We are diverting to Jakarta and expect to land in about fifteen minutes.'

Captain Moody had no sooner replaced the PA handset when they found themselves in the hazy conditions again. Once more the St Elmo's fire danced on the windscreen and their headsets crackled loudly with the static. There was no doubt that X-ray Hotel had climbed back into danger and Moody's first thought was to descend out of it as quickly as possible.

'Christ, we're not staying here.'

Moody slammed the throttles closed and lowered the nose for descent to 12,000 ft. Suddenly number two engine surged violently. It autorecovered and surged again. The bangs from the engine were so loud they could be heard on the flight deck and the entire aircraft shook. Having lost all four earlier the crew were reluctant to shut one down again themselves, but they had little choice.

'Shut down drill number two engine', shouted Moody.

The other two performed the drill and, with the fuel cut, engine power ceased. X-ray Hotel was now back on three engines. The failure of the number two engine for a second time gave the crew quite a shock, for it indicated that the damage sustained was worse than previously realised. Would another engine fail again?

'Speedbird 9,' radioed Greaves, 'we've lost another engine. We're now on three and we're descending to 12,000 feet.'

Jakarta Control approved the lower level and issued radar vectors for their prompt return to Halim.

'Ladies and Gentlemen,' spoke the captain again on the PA, 'I have had to shut down number two engine as it was rough running. We shall be landing in Jakarta in about ten minutes.'

The general feeling aboard BA 009 was the sooner on the ground the better, but it was now obvious that some gentle nursing of the aircraft was going to be required to ensure a safe landing. Rather than adjust the power and risk further upsets, Moody used the speedbrakes to control aircraft speed, and once over the mountains began a shallow descent with speedbrakes deployed. Fortunately, Jakarta weather was fine with a calm wind, little cloud and good visibility. Runway 24 was in use and the temperature was 26°C. A few minutes later, Control instructed Greaves to call Jakarta Approach on 119.7 MHz.

X-ray Hotel now approached Halim Airport from the south-east but the aircraft was too high for landing and some height needed to be lost. Moody decided to overfly the 'AL' non-directional beacon, situated on the approach to the landing runway, at about 10,000 ft (3,050 m) and then to proceed initially to the north-west before banking right in a shallow descending turn out over the sea. He would complete the let-down by flying all the way round to land to the south-west. It was a non-standard approach which suited their purposes, and they would be guided on the way by radar which would direct them onto the instrument landing system (ILS). Fortunately, the crew had been into Halim in the daylight only the previous day, so they had recent experience of the airport. They knew there was a lot of high ground in the area.

X-ray Hotel was cleared initially to 2,500 ft (760 m), but as descent continued Captain Moody was informed of another difficulty. Normally the ILS displays both runway centre line and descent path guidance, but at Halim only runway direction guidance was available. On a big jet, judging the correct descent profile for the approach to land without radio aids is difficult at the best of times, but after a major emergency it was the last thing the captain wanted. Fortunately, visual approach slope indicators (VASIs) were available, if they could see them, for in spite of the clear visibility indicated by the controller, the conditions were very misty. It was more like dust haze, in fact, and there was a sulphurous odour in the air.

As X-ray Hotel completed the turn and approached the airport from the north, heading 210° under radar control, the flight was further instructed to descend to 1,500 ft (457 m) and was cleared to pick up the ILS extended runway centre line. Halim lay to the crew's right, about eight miles (13 kilometres) away, but the view through the first officer's number two window still appeared hazy.

'Can you turn up the runway lights?' requested Greaves over the radio.

Moody turned the 747 onto final approach and suddenly, from a dark

A landing light window showing the opaque effect of the sandblasting. The 747 windshields were similarly affected. *(British Airways)*

area on the ground ahead, the brightened runway lights could be seen, but the view was very blurred. It was as if there was oil on the windscreen but the wipers and washers had no effect. Unknown to the crew, the 'cloud' which had choked the power from the engines had also sand-blasted the windscreens. As BA 009 got nearer the brightness of the runway lights on the opaque windscreen blinded their vision and they had to be turned down a little. The central and inboard sections of the pilot's windows were worst affected, but Moody managed to catch sight of the left side of the runway through a relatively clear three- or four-inch-wide outer strip. Leaning sideways he established on the runway centre line and gingerly commenced descent on the final approach.

Greaves retained the distance-measuring equipment on his side and called out the recommended descent profile of 300 ft per mile (57 m/km): distance four miles, height 1,200 ft; distance three miles, height 900 ft; etc. He switched on the landing lights but their covers were also opaque and they gave almost no illumination. Greaves then called the tower on 118.3 MHz and was given the clearance to land.

Captain Moody's view was not very clear but at least he could pick out the VASIs and had some descent slope indication. Sitting in this bent position, however, made it difficult to scan the instruments at the same time, an essential part of flying a large aircraft. As an aid, Greaves called out the heights every twenty-five feet (7.6 metres) from the radio altimeter. Townley-Freeman called the speed and the engine power settings. Cautiously X-ray Hotel groped towards the runway. On three engines, with visibility curtailed, difficulty in reading the flight instruments and no radio descent path indication, Moody had his work cut out. It was, in the immortal words of the Hampshire captain, like trying to negotiate your way up a badger's arse. The cabin crew had been instructed to prepare for an emergency landing and they waited anxiously at their stations.

BA 009 finally crossed the threshold of runway 24 at 150 knots but, as Moody prepared for touchdown, he still found his forward vision severely reduced. The lack of landing lights did not help. Normally pilots look far down the runway to judge height on landing but Moody could view only the hazy outline of the left-hand runway lights closest to him. Using all his experience he eased back gently on the control column and the wheels kissed the ground. It was a perfect touchdown. In the cabin the passengers spontaneously applauded. The time was 14:10 GMT and the emergency had lasted twenty-five minutes, thirteen of which had been totally without engine power. In a period of less than a quarter of an hour they had been to the edge of great danger and back again. Moody ran the aircraft right to the end of the 9,800 ft (2,990 m) runway, then, with Greaves's help, managed to feel his way round the turning pad and to back-track towards the terminal. Eventually they groped their way to a turn-off but as they approached the stand area the bright lights completely obscured their view.

'That's it,' said Moody, 'we're stopping here.'

Greaves called the tower to say they were unable to taxi further and asked for a towing truck to pull them to the arrival gate. Townley-Freeman then fired up the auxiliary power unit and shut down the engines. It was all over. The tension drained from their bodies and they were flooded with relief. There was no doubt it had been a close-run affair, but the cause of their predicament remained a mystery. As the arrival of the tow truck was awaited the flight crew had time to reflect on their experience. Black dust seemed to lie everywhere on the flight deck. Barry Townley-Freeman wiped his hand across a surface and rubbed the substance he picked up between his thumb and fingers. It was gritty, sooty and had a sulphurous smell.

'I reckon that's volcanic ash, you know.'

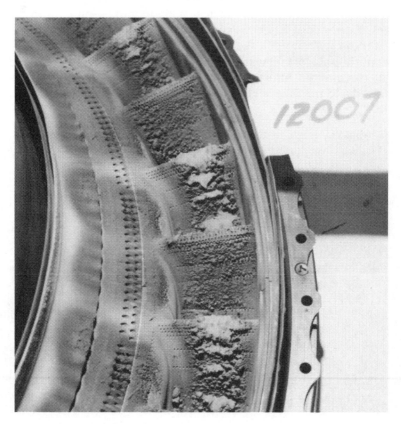

Fused volcanic ash on the number two engine high pressure nozzle guide vane. *(Rolls-Royce Ltd)*

No-one believed him at the time, but the next day he was proved right.

On the evening of 24 June, Mount Galunggung, the active volcano situated on the south Java coast 100 miles (160 kilometres) south-east of Jakarta, had erupted violently. Giant plumes of ash and grit had been hurled eight miles (thirteen kilometres) into the air. The explosion had created a vast volcanic storm of thick, hot, sulphurous gases and high electrical activity. No aviation warning was given. North-easterly winds aloft at twenty-five to thirty knots had blown the plume across the path of BA 009 and it had been engulfed in ash. Flying through the hot grit at speed had the effect of sandblasting X-ray Hotel's leading edges and had stripped paint and caused the opaque surface on the windscreens and landing light covers. The dust had penetrated aircraft sensors such as pitot tubes which sense dynamic pressure to measure airspeed. The discrepancy in the airspeeds was the result. The engine nacelles, intakes and fans were shot-blasted and stripped clean. Erosion of the compressor blades had occurred and the ash, which was of a silicate material, had fused in contact with the hot metal of the combustion chambers and turbines. Deposits of fused volcanic ash up to half an inch in diameter were later discovered in all the engine tailpipes. The effect was sufficient to disrupt the airflow and was similar to dampening a blaze with sand. As a result the engines had flamed-out and all power had been lost. At 13,500 ft (4,115 m), X-ray Hotel had broken into clean air and the least

The entire crew of BA 009, with the then deputy chairman, Mr Roy Watts, at the British Airways Award presentation ceremony. *(British Airways)*

damaged engine, number four, had roared into life. Subsequent climb to 15,000 ft (4,570 m) had taken them back into the ash cloud. It was a close escape which might have ended tragically had not the skill, coolness and persistence of the flight crew won the day. But the passengers and crew of BA 009 were lucky, too, for the aircraft only just emerged from the volcanic storm in the nick of time.

'If the base of the ash cloud had dropped to the sea,' said Moody, 'so would we.'

Captain Moody was awarded the Queen's Commendation for Valuable Service in the Air, plus the British Airline Pilots' Association Gold Medal, the Guild of Air Pilots and Navigators Hugh Gordon Burge Memorial Air Safety Award, British Airways Board Certificate of Commendation, Award as one of the twelve 'Men of the Year' in Britain, and a Lloyds of London presentation set of crystal decanters. SFO Roger Greaves was awarded the British Airline Pilots' Association Gold Medal and a Lloyds of London presentation silver tray. SEO Barry Townley-Freeman was awarded the Flight Engineers International Association's Frank Durkin Award and a Lloyds of London presentation silver tray. CSO Skinner was awarded the Queen's Commendation for Valuable Service in the Air, the British Airways Board Certificate of Commendation and a Lloyds of London presentation silver tray. Those mentioned above also received, together with the remainder of the crew, the British Airways Customer Service Award.

Of all the awards, accolades and applause received by the captain and crew of BA 009, however, no praise was more heartfelt than that offered by one of the Australian passengers. With great emotion she sobbed, 'I'll never rubbish the Poms again.'

Epilogue

Emergency proved to be a most enjoyable book to write, and during its production many interesting and gracious people were met along the way. Much kindness and help was willingly given. The fact that most participants were quite humble about their experiences, in spite of the heroic endeavours involved, served only to increase their stature. On some occasions help was forthcoming under the most extraordinary pressure, as in the case of Captain Pat Levix, who still found the strength to offer assistance only shortly after his son Jim was discovered murdered by persons unknown in 1987. Such men are gracious indeed.

The author managed to contact, sometimes with great difficulty, at least one flight crew member from each of the major events outlined in the chapters of the book. The search was conducted from Anchorage to Auckland and from Ottawa to Miami and involved some time-consuming 'detective' work. Fortunately, some lucky breaks during the process contributed to success. A few gentlemen, such as Harry Orlady of the San Francisco office of NASA's Aviation Safety Reporting System, were able to furnish information which proved invaluable to the completion of the book and the tracing of individuals. Without such help *Emergency* could not have been written.

All the participants contacted were interviewed either by letter, telephone or in person and each was given the opportunity to read the manuscript. Advice was freely given on improving and correcting the texts and the assistance was gratefully received. The final drafts, without exception, were accepted and approved by all concerned, giving *Emergency* an accuracy and authenticity not always found in aviation publications. It is hoped that the reader has been given a refreshing insight into the drills and procedures of modern flight crew members, not only during difficult and unusual circumstances, but also in the day-to-day operation of an aircraft. It is seldom as simple a task as it seems.

In all the incidents recounted in this book, in spite of the seriousness of some of the situations, not a single life was lost, and this in itself is a great credit to the aircrew involved. The accolades received by the participants in these events are a testament not only to their skills, but also to the abilities of pilots and flight engineers everywhere. The problems outlined in *Emergency* are but a sample of the difficulties, some more critical than others, faced on rare occasions by crew members in the execution of their duties. Few of these incidents are heard outside the profession, and of those who perform their tasks so successfully, most remain unknown and unsung heroes.

The men and women of the airlines who fly their machines throughout the world give of their best at all times and are ready for any contingency. When the completely unexpected arises, as in some of the stories of this book, crews are known to respond with ingenuity and resourcefulness. And in the everyday operation of an aircraft, the expertise and professionalism exercised daily by crews during flights contribute significantly to making air travel routine, commonplace and, most importantly, safe. The travelling public can feel comfortable with the knowledge that aircrews, like all other personnel within the aviation industry, are doing their utmost to maintain standards. It is hoped that *Emergency* has increased the general awareness of the capability of aircrews and that it inspires confidence. Above all, it is intended to reassure.

Abbreviations and Glossary

AC	alternating current
ADF	automatic direction finder
Aileron	wing turn control
AINS	area inertial navigation system
ALPA	Airline Pilots' Association (US)
Alternate	diversion airport
APU	auxiliary power unit
ASI	airspeed indicator
ATC	air traffic control
ATCC	air traffic control centre
ATIS	automatic terminal information service
Attitude	aircraft orientation relative to the horizontal, e.g. nose-up attitude
CAT	clear air turbulence
Cb	cumulonimbus cloud
circuit breaker	breaks electrical circuit
CSD	cabin services director
CVR	cockpit voice recorder
DC	direct current
DF	direction-finding
DME	distance-measuring equipment
Drag	air resistance to motion
Drift	wind effect
EGT	exhaust gas temperature
Elevators	tailplane climb and descent control
EPR	engine pressure ratio
ETA	estimated time of arrival
FAA	Federal Aviation Administration (US)
FDR	flight data recorder
F/E	flight engineer
FIR	flight information region
Flaps	wing trailing edge lift devices
Flare	arrest of descent on landing
Flight level	Level expressed in hundreds of feet, e.g. FL 390 = 39,000 ft

FMC	flight management computer
FMS	flight management system
F/O	first officer
'g'	effect of gravity
Gear	undercarriage
Glide path	approach descent profile
GMT	Greenwich Mean Time
Go-around	missed approach procedure
Graduated take-off	reduced power take-off
HF	high frequency (long-range radio)
Hold	holding area before approach to land
IAS	indicated airspeed
ILS	instrument landing system (runway centre line and glide path radio guidance)
INS	inertial navigation system
kHz	kilohertz
knot	nautical miles per hour
Mach number	aircraft speed relative to local speed of sound
mayday	radio distress call
mb	millibar
MEL	minimum equipment list
MHz	megahertz
NDB	non-directional radio beacon
nm	nautical miles
NTSB	National Transport Safety Board (US)
PA	public address system
psi	pounds per square inch
RMI	radio magnetic indicator
rpm	revolutions per minute
Rudder	tail fin yaw control
RVR	runway visual range
SEO	senior engineer officer
SFO	senior first officer
Slats	wing leading edge lift devices
sm	statute mile
S/O	second officer
Spoilers	speed brakes used to spoil lift
Squawk	radar transponder coded transmission
Squawk code	aircraft transponder code

Emergency

Squawk ident	press aircraft identification button
Stabiliser	variable incidence tailplane
Stack	hold
Stall	loss of lift
TAS	true airspeed
Transponder	airborne equipment which received a ground radar signal and responds by transmitting back a coded signal.
V1	take-off go or no-go decision speed
VR	rotation or lift-off speed
V2	initial safe climb-out speed
VASI	visual approach slope indicator
VHF	very high frequency (short-range radio)
VOR	VHF omni-directional radio beacon
Yaw	aircraft nose movement, left or right, in horizontal plane

Bibliography

Chapter One – Forced Entry
Unofficial reports, magazines and newspapers

Chapter Two – That Falling Feeling
UK Air Accident Investigation Branch, AAIB-AAR1/92

Chapter Three –Pacific Search
Air New Zealand Staff Magazine

Chapter Four – The Windsor Incident
National Transport Safety Board Report, NTSB-AAR/73/2
Eddy, Paul; Potter, Elaine and Page, Bruce, *Destination Disaster*, Hart,
    Davies, McGibbon, 1976
Godson, John, *The Rise and Fall of the DC-10*, New English Library, 1975

Chapter Five – Don't be Fuelish
Pan Am Incident Report
Canadian Air Safety Board Report, T22–64/1985E

Chapter Six – The Blackest Day
Arey, James A., *The Sky Pirates*, Charles, Scribners' Sons, 1972, Ian
    Allen, UK, 1973
Redfield, Holland L., *Thirty-five years at the outer marker*, Pitot
    Publishing Co., 1981

Chapter Seven – Ice Cool
National Transport Safety Board Report, NTSB-AAR-82/4

Chapter Eight – Roll Out the Barrel
National Transport Safety Board Report, NTSB-AAR-86/3
National Transport Safety Board Report, NTSB-AAR-81/8
AL PA Report, 1981

Chapter Nine – Strange Encounter
Tootell, Betty, *All Four Engines Have Failed*, André Deutsch, 1985
Diamond, Captain Jack, 'Down to a sunless sea', BALPA 'Log' article,
    April, 1986

Index

Page numbers in *italics* refer to illustrations.